media
MANUAL

The Continuity Handbook

Third edition

m
media
MANUAL

The Continuity Handbook

A guide for single-camera shooting

Third Edition

Avril Rowlands

Illustrated by Colin Cant

Focal Press
An imprint of Butterworth-Heinemann Ltd
Linacre House, Jordan Hill, Oxford OX2 8DP

ℛ A member of the Reed Elsevier plc group

OXFORD LONDON BOSTON
MUNICH NEW DELHI SINGAPORE SYDNEY
TOKYO TORONTO WELLINGTON

First published as *Script Continuity and the
Production Secretary in Film and TV* 1977
Second edition 1989
Third edition 1994

British Library Cataloguing in Publication Data
Rowlands, Avril
   Continuity Handbook – 3Rev.ed. – (Media Manual Series)
   I. Title II. Cant, Colin III. Series
   791.430233

ISBN 0 240 51391 6

Library of Congress Cataloguing in Publication Data
Rowlands, Avril.
   The continuity handbook/Avril Rowlands. – 3rd ed.
   p. cm. – (Media manuals)
   Rev. ed. of: Continuity in film and video.
   ISBN 0 240 51391 6
   1. Motion pictures – Production and direction. 2. Video
recordings – Production and direction. 3. Continuity (Motion
pictures, television, etc.) I. Rowlands, Avril. Continuity in
film and video. II. Title. III. Series.
PN1995.9.P7R67
791.43'0233–dc20

94–27790
CIP

Printed and bound in Great Britain by
Biddles Ltd, Guildford and King's Lynn

Contents

Preface

Since publication of the second edition of my book, *Script Continuity and the Production Secretary* in 1989, the revolution which television was undergoing has, if anything, quickened in pace. The industry has witnessed, and is still witnessing, technological change on a massive scale, and allied to it has been a major reshaping of broadcasting in this country. Driven by these two revolutions, those of us who have been associated with broadcasting for many years can sometimes barely recognise the industry in which we started. As with all changes there are pluses and minuses. Many established practices have been swept away, and new styles of working as well as new techniques and terminology have had to be learnt.

What has remained virtually unchanged however, is the role of continuity, although the importance of specialised observation as being central to single-camera shooting is increasingly under attack in a world where costs tend to over-ride all other considerations. One result has been that, other than in the crewing of major drama series, there is a tendency to drop the PA/Continuity from location shoots. This is, to my mind, a false economy which invariably leads to more costs being incurred in post production, while the editor searches the rushes for the shot required, as well as a poorer finished product as the effects of the lack of day-by-day continuity become apparent.

In 1989, the increasing sophistication of video, especially in the area of post production, led to a wider use of videotape in single-camera shooting in a way undreamt of a decade before. This is still true although drama is still mainly shot on film. With HDTV and Pal Plus on the horizon, wide-screen television has become something more than a twinkle in the manufacturer's eye, and many dramas are being shot on Super 16mm film for their 'bankability' value. Just to confuse the issue, however, post production of these programmes is often done on videotape, frequently using non-linear systems.

So although the role of continuity remains essentially the same, it is necessary for the person doing the job to have a grasp of these latest developments in order that he/she can provide the correct information for the editor.

This new edition of the book includes chapters on these latest developments. However, as I wrote in the Preface to the second edition of my book *The Television PA's Handbook*, by the time this revised book reaches the shops it will already be outdated in some respects. That is the nature of the speed of change in the industry.

Inkberrow,
Worcestershire

Acknowledgements

I would like to thank BBC Pebble Mill Post Production Unit for allowing me to use their own dubbing cue sheet in my book. I would also like to thank Keith Schofield for his kindness in writing the chapter on the technical information required by cameramen, and I am, as ever, indebted to my husband, Christopher, not only for the chapters on post production, but also for his unfailing advice and support during the writing of each edition of this book.

Introduction

There you stand, dressed in your oldest jeans and thickest sweater (unless you're unbelievably lucky and start in fine weather): a stopwatch resting imposingly upon your bosom—maybe a Polaroid camera also—and a clipboard clutched as a lifeline in front of you. Pencils, ready sharpened, are secreted somewhere about your person; virginal continuity notes and unmarked pages of script lie neatly gathered under the clip on top of your clipboard, and securely tethered (in case of high winds) with a large rubber band. You have left your typewriter perched on a couple of packing cases in the cowshed nearby. You may, being prepared for any eventualities, have a sheet of plastic covering the paperwork. In your capacious waterproof bag somewhere near your feet are spare copies of the script and the schedule, together with spare pens, pencils, adhesive tape, scissors—spares of everything, in fact, even down to the odd aspirin and safety pin.

Around you are the rest of the unit, all equally uncomfortable despite the 'getting it together' drinks at the director's expense the previous evening. Everyone is bursting with the effort of being alert and appearing as impressive, busy and efficient as the others. One of the electricians cracks the first joke; the clapperboy comes up to you—or you go up to him—and he pledges to keep you informed of changes of camera roll. You, in turn, promise to keep an eagle eye on the slate numbers, and if he chalks up the wrong number, not to hold him up to general ridicule—at least, not in the first week of shooting. He says, not trying to be original, 'How about starting with Slate 1?' and you duly laugh and agree.

The cameraman, alternately gazing with a worried frown at the sky and then staring at his exposure meter and shaking it with incredulity, stops to ask the director: 'What's the first set-up?' The director interrupts his agitated first assistant, who is engaged in an earnest explanation of the lack of coffee/rolls/sausages and rushes over to his cameraman (who must be humoured). The first assistant has a go at the second assistant about the lack of coffee/rolls/sausages and the second assistant sends the third assistant off on a search. You become unwittingly involved with the boom operator who, under the guise of telling you the sound roll number—which, believe it or not, is roll one—starts to get too familiar. This does nothing for your image on the first day of filming, so you leap off to the director and look concerned and interested while he explains his first shot to the camera crew.

Artists have now been called for, and appear in a flurry of make-up assistants with large powder puffs and dressers with large safety pins. The director and first assistant are smothered in a round of complaints and embracings and the air is full of expressions in vogue with top artists. The rest of the unit stand round looking rather bored, except for the electrician who is looking round for likely talent.

Rehearsals commence, and you write down everything from the colour of the grass to the markings on the third rock on the left of frame. As rehearsals continue you neatly rub out what you first wrote as moves are changed, and then you rub out the second lot of notes as things change again.

Finally, all is ready for a take. The first assistant calls for quiet, the director calls 'turn over', the sound recordist calls 'speed', the camera operator 'mark it', and the clapperboy, who's been itching to do his bit with the clapperboard for the last minute, claps it and jumps out of the way. You, having already written the scene, the slate, the description of the shot, the set-up and the artists, are poised, one hand on the stopwatch, the other holding a razor sharp pencil.

The director calls 'action' and you're away. He almost instantly calls 'cut' due to heavy artillery fire from the army practice range two miles off. (The army wouldn't play ball and stop their activities for the filming.) The director casts a pained glance at the first assistant (who couldn't persuade the army); the first assistant casts a pained glance at the second assistant who hurries up to the third assistant (who has just arrived hotfoot and panting with the coffee, etc.). The clapperboy painstakingly rubs out '1' and writes '2' on his board, with a stick of chalk embedded in a small piece of foam, attached to the clapperboard by a long clean piece of string. By the end of the shooting, the string is so well worn it has snapped off and the 'hunt the chalk' game takes place hourly.

You write 'NG, noise' on your notes—the board is clapped, the army silenced, the gaffer electrician stops work in protest as the coffee/rolls/sausages meant for the crew have been used to bribe the army into silence, and your first day doing the job of continuity has started.

But what, in fact, is the job and how does one do it?

Note

Although I wanted to avoid any sexual stereotypes in this book, as there is not a suitable pronoun referring equally to 'he' and 'she', I have referred to the person doing the job of continuity as 'she'. I hope I will be forgiven for that or for any other apparent stereotypes. They are not intended.

Dressed sensibly and warmly and hung about with the trappings of the trade—clipboard, Polaroid camera, stopwatch, large canvas filming bag, etc.—you stand with eyes alert and pencil poised.

3

Out of Sequence Shooting: 1

When a film or video is being made it is not shot in a consecutive manner according to the story. That is, there is no progression from Scene 1 on Day 1 to Scene 25 on Day 25 or whenever shooting stops. It is generally shot out of sequence.

Take, for example, the following story:

Story

Two guards are on the battlements of a castle. It is daytime. They see a dusty rider on the road, galloping towards them. One guard reaches for his gun. The rider suddenly reins in his horse and looks up at the castle. We see the rider's view of the castle.

Cut to a shot of a woman on board ship.

Back to the battlements where one guard is keeping watch on the rider and the other guard runs down some stairs to raise the alarm.

Cut again to the woman on the ship.

Cut to a room in the castle, full of drunken guards. They hear the alarm, leap up and rush from the room.

In order to plan when you are going to shoot each separate piece, this story would be broken down initially into scenes:

Scenes in story order

Scene	Subject	Int./Ext.	Time
1.	Castle battlements	Exterior	Day
2.	Dusty rider on road (seen from castle)	„	„
3.	Castle battlements	„	„
4.	Dusty rider (from castle)	„	„
5.	Rider's viewpoint of castle	„	„
6.	Woman on ship	„	„
7.	Castle battlements. One guard exits Camera follows him down stairs	„	„
8.	Battlements. Guard raises gun	„	„
9.	Woman on ship	„	„
10.	Mess room of castle	Interior	Day

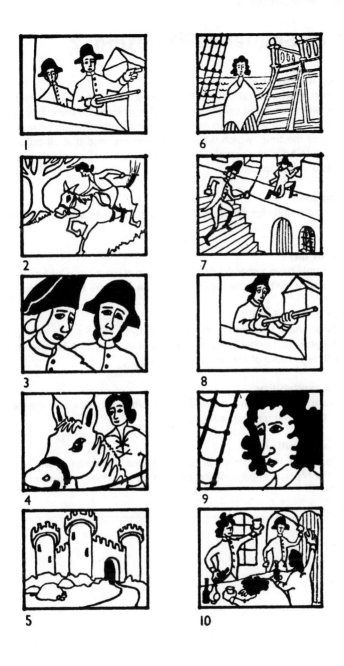

Scenes in story order:
1. Castle battlements; 2. Dusty rider; 3. Battlements; 4. Rider; 5. Rider's POV of castle;
6. Woman on ship; 7. Battlements. Guard exits; 8. Guard raises gun; 9. Woman on
ship; 10. Mess room of castle.

Out of Sequence Shooting: 2

Shooting in story order

If the shooting were to take place in that order there would be a great deal of time-consuming trudging up and down flights of steps for the whole unit with all the equipment. It would also mean that all the artists would be required for the whole of the time, which would be expensive, and that all the locations would have to be in the same area, which is unlikely. So the script is broken down into a shooting plan.

Shooting plan

The plan or schedule for shooting is worked out bearing a number of things in mind.

Number of days

You might only have a specific number of days or weeks in which to shoot the story. You must try to fit the schedule into the requisite time allotted.

Position of locations

It takes time and a fair amount of organisation to move a unit from place to place. Therefore it is better to shoot everything in location A before moving to location B, regardless of the order in the script.

Location availability

If a particular location is only available on certain dates, the rest of the locations will have to be arranged accordingly.

Actor's availability

As with locations, if a particular artist is engaged, the schedule will have to fit in with their prior engagements depending on their importance to the production.

Shooting schedule

Day	Scene	Location	Interior/ exterior	Time	Characters
1	6	Ship	Ext.	Day	Woman
	9	Ship	Ext.	Day	Woman
2 a.m.		*Unit move to castle location*			
p.m.					
	1	Battlements of castle	Ext.	Day	2 guards
	3	Battlements of castle	Ext.	Day	2 guards
3	8	Battlements	Ext.	Day	1 guard
	7	Battlements and steps	Ext.	Day	2 guards
	2	Road (seen from battlements)	Ext.	Day	Rider and Horse
	4	Road (seen from battlements)	Ext.	Day	Rider and Horse
4	5	Road—LS castle	Ext.	Day	-
	10	Mess room	Int.	Day	20 soldiers

Shooting order
A shooting order like this allows for the fact that the ship is at a different location from the castle. It uses the artists economically (the woman is only engaged for one day's shooting), and the crew are not required to trudge up the steps to the battlements more than once, which allows a much faster shooting time.

Shots in Scenes Out of Order

A film or video is not only shot out of sequence but the shots within each scene are generally not taken in order.

To save time
This happens for a number of reasons, all to do with wasting as little time, and consequently money, as possible.

Usually only one camera is used and it has to be moved when different angles are shot. It is far faster to take all the shots from one particular area, whether they are from the beginning, middle or end of the sequence. The camera does not have to be moved around too much between shots and the lighting needs minimum adjustment until a major re-light is required for a totally different angle. Also props do not have to be constantly moved here and there to keep them out of the way of the camera, the lights and the vast army of people hiding behind the camera.

Part of a set
This set has been lit and the camera set up to take shots from one angle. Much time
will be saved by taking all the shots needed from this area at the same time,
irrespective of where they occur in the scene, rather than moving the equipment and
lights from one side of the set to the other for each shot.

The Work of Continuity

The person responsible for continuity provides a complete, written document of the shooting. It should be a very detailed and accurate yet very concise work. By concise, I mean that while a two-page essay on each shot in a flowing literary style may be aesthetically pleasing it will be virtually useless as a working document, as no one will have the time to read it for the required information.

Use of document: (A) to unit on location
As the material is shot out of sequence it is obviously necessary to have some record, built up over the shooting period, of what was actually shot in order to preserve continuity from one shot to another and from scene to scene, irrespective of the shooting order.

This means that the closest attention must be paid to every shot. A description of the shot, how it begins and ends, the relationship between the characters' actions to the dialogue, the position of props—in fact, as much of what the camera sees as possible—must be written down for immediate and later reference.

You might be asked by the cameraman to note down certain technical information, such as the lens, the stop, distance and special filters. This would assist him to match shots or compose retakes.

Use of document: (B) to the film or videotape editor
The document also contains certain technical information which is of interest only to the editor. Editors do not want to know details of costume and props, as such notes are of use only while shooting is still taking place. But they are interested in the reference point for the shot—the slate number in film or the 'in' point of timecode in video—the shot description, the agreed take and, most especially, the reason why a take was considered NG.

Back to the original story
The role of continuity ensures that when the finished material is assembled and finally shown it will flow in chronological order in a smooth way, and that continuity is preserved in action, sound, costume and props within each scene and from one scene to another.

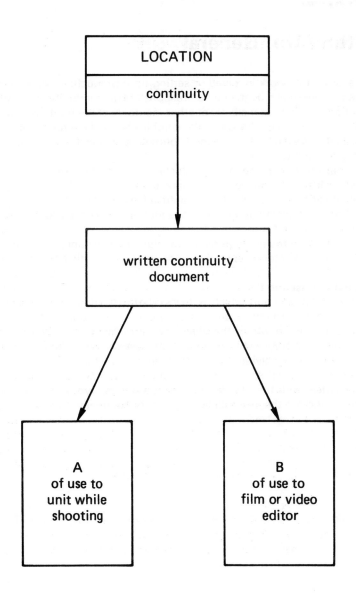

```
┌─────────────────────────┐
│        LOCATION         │
├─────────────────────────┤
│       continuity        │
└─────────────────────────┘
             │
             ▼
┌─────────────────────────┐
│   written continuity    │
│       document          │
└─────────────────────────┘
       ╱           ╲
      ▼             ▼
┌──────────┐   ┌──────────┐
│    A     │   │    B     │
│ of use to│   │ of use to│
│unit while│   │film or video│
│ shooting │   │  editor  │
└──────────┘   └──────────┘
```

Continuity work
You provide a written record of the shooting. This document is built up while shooting takes place and is used as a reference. It is also of interest to the film or video editor for the technical information it contains.

Setting Up: General

There is a lot of work involved in setting up a production. Much of this might be undertaken by the person responsible for continuity on location, or it might be the responsibility of other members of the production team. A lot depends upon the size and complexity of the production, past precedents, the size of the overall budget, and the type of company making the film or video.

It is hard to generalise, as a production can employ a whole army of people or just one dogsbody who does everything—and that job can be the most satisfying! It can also be confusing as someone you are used to calling a First Assistant might well masquerade under the title Production Manager in your next job!

The setting up you are required to do might involve none, some, or all of the following chapters, depending upon your own particular job.

Scripted or unscripted
The type of production which is to be shot will, of course, vary enormously, ranging from the multi-million pound cinema epic to the video of the board meeting where the chairman gives his annual address. The requirement for continuity is there in both cases but to differing degrees. In this book I have divided productions into those that are scripted and those that are unscripted—drama and documentary types. The following chapters deal with the pre-shooting period of both types of production. But first, who are the key people in the production team?

The life of the 'dogsbody' can be interesting and rewarding.

The Production Team: 1

The number of people on the production team will of course vary according to the complexity of the production and the size of the budget. But the following gives some idea of the names and responsibilities of some of the key people. Don't be put off by the plethora of different and sometimes confusing names for people doing, in essence, the same job.

Producer
The producer might be responsible for commissioning the scripts, engaging the director and overseeing the casting. He/she has overall responsibility for making the production within the allotted budget and to the time-scale agreed.

Production Associate
The production associate is responsible for the overall budget and for booking basic facilities. They might have Finance Assistants working to them.

Director
The director's primary responsibility is to the script. He/she is responsible for turning the printed page into the visual medium of television. The job requires a good understanding of the mechanics of television production together with qualities of leadership and the ability to communicate his/her creative interpretation of the script into a finished work.

First Assistant
Also can be known as the Production Manager or Unit Manager or Floor Manager (FM). This person is, in effect, the deputy director, working closely with the director and artists. The first assistant usually compiles the schedule of shooting, directs extras, gives call times and might be called upon to direct a second unit.

Second Assistant
Might also be known as Stage Manager (SM) or Assistant Floor Manager (AFM). The second assistant works to the first assistant, but is primarily responsible for action props and prompting and cueing artists.

Third Assistant
Also known as Floor Assistant, Call Boy/Girl, Runner. The third assistant also works to the first assistant and is responsible for calling the artists when needed and carrying out any other function assigned to them.

Don't be put off by the plethora of different and sometimes confusing names for people doing in essence the same job.

The Production Team: 2

Production co-ordinator, Production Assistant, Script Supervisor, Production Secretary
In television, the co-ordinating work in setting up a production might be done by a Production Assistant (PA), who could also be responsible for continuity, or the work might be carried out by a Production Secretary. (The additional multi-camera studio skills done by the PA working in television are covered in my other book, *The Television PA's Handbook.*)

In the feature film industry the setting up is generally done by the Production Co-ordinator, while continuity is done by a Script Supervisor. However the differences between the feature film and television industry are daily growing less and many television productions split the responsibilities for setting up along feature film lines.

Other members of the production team have self-explanatory titles:

Location Manager
Responsible for finding the locations.

Production Buyer
Responsible for buying or hiring the stage properties.

Casting Director
Assists the director with casting.

Costume and Make-up Supervisor
Responsible for costume and make-up. Often with assistants or dressers working to them.

Scenic Designer
Responsible for set design.

However few or many people form the production team, they should work closely together. One of the most important jobs undertaken by the PA during this period is to ensure that communication between members of the team does not break down and that information is passed on quickly and accurately.

One of the most important jobs undertaken by the PA during this period is to ensure that communication between members of the team does not break down

The Script

The first job in setting up—if you are working on a drama-type production—is to take a dog-eared, much-scribbled-over, almost indecipherable draft script and turn it into a nicely typed legible rehearsal script for everyone's use. This might not be the final, definitive version, but everyone wants something to work from as soon as possible.

Numbering

Every production is broken down into its smallest component—the script is divided into scenes and these, in turn, are separated into shots. It is therefore vital to number and label everything clearly.

Put the scene number and page number (i.e. Sc. II (13)) at the top and bottom right-hand side of each page. If there are a number of episodes you might find it useful to incorporate the episode number with the scene number (i.e. Episode 2, Scene 1 becomes 201; Episode 3, Scene 12 becomes 312, etc.).

Colour-coding

Colour-code different episodes by having each script photocopied on different colour paper.

Scenes

A change of location denotes a new scene. Always start a new scene on a fresh page, even if it is only one line.

Body of script

The script you type is a working document. You will be adding continuity notes to it during shooting and others will also require plenty of space for their own requirements. Type the script on the right-hand side of the page leaving a wide margin on the left; type stage directions in capital letters and dialogue in upper and lower case; and type dialogue in double space so that alterations can easily be made. Try not to carry over a piece of dialogue from one page to the next. The script should be typed single-sided only.

Size of script

Scripts are normally printed on A4 size paper, but it is very helpful to artists and crew to have copies printed on A5 paper for use on location. These smaller versions are far easier to handle—to keep in coat pockets for example—than the larger size, especially if a number of episodes are involved.

SCENE 505 : EXT. STREET

PAULINE WALKING

SCENE 506 : DAY. INT. OFFICE

LAURA IS SITTING BEHIND HER DESK, WHICH IS LITTERED WITH PAPER AND PHOTOGRAPHS. EVEN THE CHAIR IS FULL OF PAPERS. THERE IS A KNOCK ON THE DOOR.

LAURA

Come in!

PAULINE ENTERS THE ROOM AND APPROACHES THE DESK

PAULINE

I've come about the advertisement.

LAURA

Just a minute. (SHE FINISHES WHAT SHE IS DOING BEFORE LOOKING UP) You've come about...?

PAULINE

The advertisement. You know, the one in the paper.

LAURA

Oh, of course! (SHE SHAKES HANDS) Do sit down. Just put that junk anywhere. I always work in an awful mess, don't you? Now you're...?

Example of script layout

Script Breakdown

After the script has been typed and distributed, it is advisable to type a breakdown of the scenes. Such a breakdown will assist in working out the schedule of shooting and also prove useful to the set designer, costume and make-up, the property manager and many others. The breakdown should contain:

Episode Number: if there is more than one episode

Page Number: of rehearsal script

Scene Number: every change of location is given a fresh number

Interior/Exterior: is the scene to take place inside or outside?

Day/Night: it might be helpful to continuity and important to the script to write the actual day and even the time. For example, 'Day 1', 'Night 1', 'Day 2', or 'Day 1, 10.00 a.m.', 'Night 1, 8.00 p.m.'

Location: give the location, e.g. 'garage', or the actual address if that is known

Characters: list all characters required in the scene, together with details of supporting artists

Synopsis: it is very helpful to give a brief synopsis of the scene. This enables production people to work from the breakdown without having to refer to the script when planning schedules, costume and make-up changes and so on.

Continuity: you might find it useful to list whether there is direct or indirect continuity between scenes or characters. It will provide a guide for costume, make-up and props and be of use to you on location.

Props/Special Equipment: if you have room you could leave space for a props list or for special technical equipment.

TITLE:

PAGE NO.	SCENE NO:	INT/EXT	DAY/NIGHT	LOCATION	CHARACTERS	SYNOPSIS	CONT.
1	1	Int	Day	Railway Station	Louis Alice Porter (W.On) Passengers	She meets him off train. They arrange meeting in garden	No direct
3	2	Ext	Day	Gravel pit	Ch.Insp.Bartley James Wilcox Policemen	Police find body	No direct
4	3	Int	Night	Dining Room - Manner's House	Alice / John James Wilcox Mary Wilcox Elwyn Brand Grant Tyler Sarah Glyn	Dinner party where Alice finds that Tew has been murdered. She is regarded with suspicion by James. Sarah & John exchange glances	Alice – direct cont. with Sc. 4 Everyone else - cont. with Sc. 5
10	4	Ext.	Night	Garden (by summerhouse)	Alice Louis	She tells him of her fears. He dismisses them as irrational.	Alice – back cont. with Sc. 3 – forward cont. with Sc. 5
13	5	Int.	Night	Dining Room - Manner's House	John / James Mary / Elwyn Grant / Sarah Alice	Dinner party cont. Alice rejoins. James accuses her and Alice faints	Everyone with Sc. 3 Alice with Sc. 4

Script breakdown.

21

Timing the Script

Once the script has been finalised—for the time being at any rate—you should time it, noting down, scene by scene, the approximate durations.

You do this by reading it through to yourself, out loud, acting out the actions as they occur, i.e. 'goes to door'. Some scenes, of course, might prove rather difficult to act out, even in the privacy of your office or home. You will have to make an intelligent guess as to the length of those scenes.

Another guideline you can use is that 50 pages of script typed in the manner stated earlier and containing a mixture of dialogue and action will be approximately 30 minutes of final running time.

Time chart
Having timed the scenes, note them in chart form, adding up each scene to give a cumulative running time.

Read-through
The cast might be asked to attend a preliminary read-through of the script prior to shooting. This will be a good opportunity for you to get a more accurate timing.

Scene	Duration	Cumulative
1. Int. Living room	1.20	
2. Ext. Garden	2.00	3.00
3. Ext. Road	30	3.30
4. Int. Living room	4.20	7.50
5. Int. Hall	15	8.05
6. Ext. Car	40	8.45
7. Ext. Shopping City	1.10	9.55
8. Int. Shop (s'market)	3.05	13.00
9. Ext. S'ing City	45	13.45
10. Int. Shop (stationers)	25	14.10
11. Ext. Alleyway	30	14.40
12. Ext. Street	20	15.00

Example of time chart.

Artists and How to Book Them

You need to book artists quickly as the whole process takes some time to complete.

Artists
Artists come in all shapes and sizes and to suit all requirements. They are found by the producer and/or director, who either have definite ideas of what they want, or sit closeted for hours with directory and cast list laboriously trying to make a list of people to audition.

Agents
Artists are looked after by agents, who find them work, look after their finances and generally administer to their needs. Some agents tend to specialise in actors with particular skills or abilities, others cater for all requirements. They can be pleasant and helpful or downright bloody-minded. They should be approached with caution.

Auditions
You might have to circulate the agencies with details of the forthcoming production and a brief description of the cast needed. The agents then contact you with lists of people who they think suitable. After much discussion, you are given a shortlist of people to find out about. With this shortlist you ring the agencies and talk in deliberately vague and general terms in order to ascertain the availability of the person you are interested in for the dates necessary. Do not be specific because a verbal contract can be legally binding even if given over the telephone and without witnesses.

Auditions then take place with the selected people at a time and place previously arranged. After many such auditions, much discussion and countless references to the budget, a cast is agreed upon and contracts signed.*

Names and addresses
As soon as the principal artists are booked, get their names, addresses and phone numbers without delay and circulate them to people like the costume and make-up designers, who will need to contact them.

Scripts and schedules
Let your artists have scripts and schedules in good time—always supposing they are ready.

Hours of work
When you are actually shooting, keep a note of the hours the artists work and their travel time to and from the locations, as they may be entitled to payment for working overtime.

*Pay attention to clauses in the contract relating to co-productions, repeat fees and special requirements (i.e. swimming required or hair not to be cut).

Artists and their agents
Artists come in all shapes . . . and sizes. They are looked after by agents.

Background Artists

Non-speaking artists who are booked by the score—budget permitting—to add to the verisimilitude of a production are known as background artists, atmosphere artists or crowd artists. The distinction between them, walk-ons and small parts is often extremely fine, but naturally very important to the people concerned as it has a bearing on the fees paid and the status of the artist.

Definition of a background artist
A background artist is someone who, in conjunction with others, is given general directions, e.g. 'mill about' or 'act angry', but no individual direction and has no set words. A crowd of fans at a football match might all be background artists. They might applaud and cheer as the mood takes them, but are not instructed beforehand on what to say, and do not perform any specific pre-arranged actions on cue.

Definition of a walk-on
There are three grades of walk-on, 1, 2, and 3, and arguing sometimes takes place in order to determine the level of a specific walk-on. To give some idea we'll go back to the football crowd. Out of this crowd one person might be singled out to wave a flag on cue. He would then be upgraded to walk-on 1 or 2. So would the policeman who points a warning finger at the flag waver. If the policeman were given some unimportant, unscripted words to say, he would be upgraded to walk-on 3. So would the flag-waver, if he was told to utter some appropriate epithets at the policeman. If, of course, the words are subsequently scripted, then the Walk-On is given a contract as a small part.

For further information concerning these definitions ring the Film Artists' Association, 0171 937 4567.

Where to obtain background artists and walk-ons
Background artists and walk-ons are employed by agencies and can be ordered in bulk. You can ask for 50 peasants, assorted, 20 female, 30 male, between the ages of 25 and 50, or be more specific in your requirements.

There are agencies that specialise in certain types, for example, there is one which caters for stunt men, heavies and fall guys in general, and will produce a fine range of villainous-looking characters at a moment's notice.

Don't get carried away!
It can be a fairly heady business, ordering vast quantities of people, and one must be careful not to get carried away by the experience and over-order. You must also be careful not to make mistakes because although background artists are not costly individually, *en masse* the price can be formidable, especially if overtime work is incurred.

Suitable transport must be arranged and clear instructions on rendezvous points given. It is usually necessary for them to be 'chaperoned' constantly, and generally someone is detailed to keep an eye on them and have them at the right place at the right time.

Always check and double check where background artists are concerned. It is so easy to lose the odd one or two.

Background artists

1. A background artist is a human prop

2. If he or she performs an action or says a few unscripted words he or she then becomes a WALK ON.

Children and Animals: 1

If at all possible, avoid children and animals. They are complicated to book, have an enormous number of rules and restrictions attached to their use and need a lot of looking after—even though they come with a retinue of chaperones, tutors, trainers and handlers. Often their minders can prove more of a problem to organise than they are.

Children and animals can waste an immense amount of time during shooting. On balance, I prefer children because at least (if they are old enough) they understand what you are saying—but trying to make an animal do something on cue requires superhuman effort and patience—and a flexible shooting schedule.

Where to find children
Children come from two sources, stage schools and non-stage schools. Much can be said for and against both.

Some directors never seek children at stage schools, preferring to make the rounds of ordinary schools and engage the help of drama organisers. Others cast solely from stage schools as they feel that the children will have a certain knowledge of the business and sense of discipline to start with.

Wherever they come from, in the UK if they are under 16 they need licences from the local education authorities before they can appear in films.

Licences
Before the licences are granted, the authorities have to ensure that the *Children (Performances) Regulations 1968* will not be contravened. An application form for a licence has to be filled in and submitted, and this form goes into searching detail about every aspect of the shooting.

Children and animals

Children and animals can be difficult to work with. Both need more time than the usual shooting schedule can allow and vast amounts of patience.

Children and Animals: 2

Regulations

The regulations governing the use of children are very stringent, and should be examined in some detail before any individuals are booked and the shooting schedule arranged. Children's hours of work, tuition, rest periods etc. are all laid down quite categorically and must be adhered to.

When shooting, their hours must be carefully noted down as these details may be subject to inspection by the Education Authority.

Chaperones and tutors

Children under 16 need chaperones. These august ladies are to be found on any set sitting in a corner, knitting and discussing in depth how this production compares with the last, and exchanging the latest piece of gossip. They are in overall charge of the children. They must collect them from their home or their school, travel with them, keep an eye on them when shooting and make sure that they are not getting overtired or being worked for too long without a rest. They are very important to the children as they are not connected with the production and therefore act solely in the child's interest. They must also take them home, or be responsible for them in the evenings if away from home.

If the shooting occurs during school term time, tutors must be engaged to give a specified number of hours schooling each day.

Animals

Animals can be obtained from a number of places. There are a few agencies which deal in them, otherwise zoos, circuses and nature reserves are all possible places to search for the animal required.

They almost always come accompanied by a trainer or handler. The problems with animals start with their first day's shooting. They do not present any of the problems with regulations one encounters with children.

The provider of the performing animals in the UK has to comply with the *Performing Animals (Regulation) Act of 1925* with regard to training the animals.

Officers from the local authority have the right to inspect the premises with the object of preventing cruelty to animals.

1

2

Regulations for employing children and animals
There are numerous regulations governing the employment of children and animals.
They should be studied well. Any person under the age of 16 is classed as a child and
is subject to these regulations. All children must be provided with qualified
chaperones. All animals must be attended by their handler or trainer.

Locations: 1

Anywhere you choose to shoot belongs to somebody, even if it is only the local council, and you do need to get prior permission—preferably in writing. If you are filming in a street you should notify the police as you might be the unwitting cause of traffic or people congestion.

Apart from getting permission for your shooting, you might also have to arrange a suitable fee and organise indemnity against heavy-handed crews damaging the surroundings.

The director might find the locations himself or herself, or the job might fall to the location manager or the production manager. Alternatively, you might have to find the locations yourself. So how do you go about it?

How to find your location

1. *The script.* The script is obviously the first pointer. If the story is set in one of the hotter, dustier regions of Spain, it is not good trying to shoot it just outside London. You might well find a dusty region, but what about the weather?

2. *Accessibility.* It is possible to take a unit anywhere—almost. What must be considered is whether the cost and difficulties of travel to remote regions are worthwhile for guaranteed quiet, good weather and no one officiously demanding a permit.

A large production, complete with artists, make-up, wardrobe, props, lighting, camera crew, etc., is a very cumbersome body to move around. There is usually heavy equipment, large vehicles and a great number of people.

3. *Communications.* This leads on from the last point. It is necessary to have fairly fast, reliable communications with civilisation, and there must either be a means of transport available or special transport provided, often at great expense.

4. *Facilities.* You might get all the peace and quiet you need in Outer Mongolia, but what about the hotels, the lack of electricity, telephones, fax machines, or even the availability of public conveniences? This is to say nothing of evening entertainment, although a good unit will always provide its own in one way or another.

Hunting for locations

Many things must be taken into consideration when looking for locations—not just peace and quiet. Reasonable accessibility to civilisation is one factor as a unit is usually quite a cumbersome body to move around.

Locations: 2

5. *Weather*. This is a difficulty that haunts many location-hunters in this and other countries. It might be that time wasted in waiting for the weather to settle could be better spent in transporting everyone to a country with a more settled climate. In most cases it is not important to have especially good weather but, for continuity purposes, the weather should be consistent in scenes that are to be intercut.

When the schedule of shooting is compiled, it is important to build 'wet weather cover' into the schedule if at all possible, which means having stand-by interior scenes in case it is not possible to shoot outside.

6. *Noise*. This can be another problem. These days you have to travel further and further away from population centres to obtain real silence and it is amazing how noisy the heart of the country can be, what with birds, insects, tractors, animals, aircraft, etc.

Never allow mobile phones near where shooting is taking place as their constant ringing will drive the director and crew to distraction.

7. *Friendliness of the locals*. You may have found the ideal spot—it is reasonably quiet, the weather is good, the scenery just right and there are some hotels up the road, but when you arrive, you are met by an assortment of hostile locals brandishing umbrellas, rolling pins or other weapons, demanding:

    (a) that you leave immediately,
    (b) that you agree to pay large sums of money as 'protection',
    (c) that the crowd are all given parts—speaking ones of course!

Locations: other aspects

1. Weather is an important factor. A run of bad weather can prolong a shooting schedule alarmingly.

1

2. Noise is another problem to be aware of if you are shooting sync sound.

2

3. Friendliness of the locals—it helps ease the strain.

3

Accommodation

Booking suitable accommodation is often a very complicated business. It is essential to take time and trouble over it and to check and recheck the situation constantly.

It is better to overbook, initially, rather than omit to make bookings for people, as no-one would take kindly to your muttered 'I'm sorry, I forgot', if they have no bed for the night. But do not forget to cancel the rooms not needed, otherwise there will be an unnecessarily heavy charge.

Finding what is available
Get as much information as you can about the available accommodation in the area where you will be shooting. Check whether you have to contend with the holiday season, the local carnival or the annual pole vaulting contest. If there is likely to be a shortage, make a block booking quickly and sort out the details later.

Finding the hotels
It is often not easy to find hotels close enough to the location that are satisfactory for price and number and size of rooms, and that can also meet all the varied requirements of such a varied group of people.

Making the booking
If it is a large booking, or made for a long period, many hotels come to some arrangement over a reduction in price.

Find out who wants what—for example, single rooms, double rooms, etc. and try to arrange for these preferences to be upheld. It is not always possible, but if everyone is happy in their accommodation, it makes a lot of difference to the spirits and morale of the whole unit.

Some of the unit may well prefer to make their own arrangements. So long as this is acceptable to the director, they should not be forced to accept the booking you have made for them just because it simplifies things for you.

Your own booking
Make sure that you have adequate room in which to spread yourself and your mini office.

Where to sleep everyone
Trying to find accommodation to suit everyone can be difficult. Apart from the
practicalities of booking enough beds on the requisite nights in reasonable hotels you
always get problems with some members of the unit who will never be satisfied with
your arrangements.

Travel

It is very difficult to give specific details of what transport you will need and when. It depends so much on the locations, the numbers of people involved, etc.

All you can really do is ensure that every single person on the unit has some means of transport to the location—and that you have adequate transport while on location.

Overseas shooting

If shooting is to take place overseas, you should ensure that all travel arrangements are made in good time as air and shipping lines tend to get booked up. Check passports, permits and visas. Contact the press office of the Embassy or High Commission of the country in which you are to work. Arrange the necessary travellers' cheques and local currency.

You will need to make sure that correct vaccinations or inoculations are obtained for all the crew if these are required, and take out health and personal effects insurance for the production team.

If a British crew is booked, an important part of the P.A.'s job will be in the preparation of the 'carnet'. This is a detailed list for HM Customs of all stock and equipment being taken abroad. A number of copies are required and pro forma invoices for consumables, i.e. tapes and batteries, are also necessary.

When booking flights, bear in mind that you will probably be carrying a good deal of excess baggage. Sometimes deals can be arranged with airlines. It might be necessary to hire a shipping agent to meet you and help clearance through Customs.

Travel in this country

Make sure there is more than adequate transport. If 15 people are to be collected from Point A and taken to Point B, book something slightly larger than a 15-seater coach as there may always be one or two extra people who turn up at the last moment.

In arranging transport you often have to hire all kinds of vehicles, fix rail tickets, and most important, give clear rendezvous points and times for transport and people.

Find out the local taxi or car hire firm in the area where you are shooting and keep a note of their phone number.

Car parking

Make sure that there is enough car parking space available at the hotel and at the location—or find out where vehicles can be left, not too far from the location.

Transport
Arrange suitable transport for everyone whatever the location, or make sure they have equally suitable means of transport of their own.

Food (or to location cater or not . . .)

After accommodation and travel, food is one of the priorities. A number of firms specialise in feeding hungry units, from around 20 to unlimited numbers of people.

They operate from anything from converted buses, converted coaches and old lorries to the latest streamlined, fully equipped modern kitchen-on-wheels.

I cannot answer for all of them, but the ones I have experienced have all provided an excellent service—often rather too good in respect of one's waistline.

They are not cheap, but they do provide coffee and hot snacks in the morning, a full hot lunch, and tea and cakes and biscuits in the afternoon. And the time saved in having hot food ready to hand, rather than hunting around for pubs and cafes, is well worth it. The only thing lacking can be a suitable place for consuming the food, and the spectacle of members of a unit trying to shield their steaming plates from the elements—usually rain—while making a dash from the caterers' van to the nearest cover must have amused many a local inhabitant.

Other types of catering

If location catering is not to be provided, then find out in advance what restaurants and cafes are in the vicinity. See whether any of them could cater for the numbers and type of meal you require and make advance arrangements with them.

Do not forget that while *your* lunchtime meal may consist of a piece of crispbread and some cheese, many people on the unit demand something far more substantial, especially after a long, cold morning shooting in the open.

Other meal breaks

If your shooting schedule covers other meals, i.e. breakfast or dinner, make similar arrangements with a local restaurant. If you have no caterers with you, arrange to take flasks of coffee and tea for mid-morning and afternoon, and do not forget the paper cups, spoons and sugar.

If you are far from civilisation

There may be times when you find yourself shooting in remote areas with no food available from any source. There is only one answer and that is to carry your own packed lunch. This is not popular with units and if you are forced to provide sandwiches made the night before by your hotel, for consumption by a hungry crew, do at least take flasks of hot soup and coffee with you.

Location catering
Location caterers operate in a variety of vehicles and produce nourishing meals. They are usually very good value for the price.
The difficulty comes in trying to eat this tasty food while perched on the bonnet of a car doubled up to keep the rain from reducing your food to a sodden mess.

Insurance

It might be your responsibility to arrange insurance for the production. Go to one of the insurance agencies who specialise in film and television work and take their advice for it is essential that your production is properly insured. The types of insurance you might need would be:

1. *Employer's liability.* This means an indemnity against liability at law to pay compensation and claimants' costs and expenses in respect of injury to any business employee arising out of and in the course of his/her employment by your company in connection with the business caused during the period of insurance and within the territorial limits.

2. *Public liability.* Means an indemnity against liability at law to pay compensation and claimants' costs and expenses in respect of:

(a) accidental injury to any person
(b) accidental loss of or accidental damage to property occurring during the period of insurance and in connection with the business within the territorial limits.

3. *Special insurance cover for artists.* Equity will provide a special insurance to cover specific artists if the nature of the work means that such an insurance is asked for by the artists' agent.

4. *Insurance for re-shoot.* Insurance to cover a re-shoot in the event of fire, accident, theft, loss or damage to the recorded tapes or shot film.

Insurance for a reshoot
Insurance to cover a reshoot in the event of fire, accident, theft, loss or damage to the recorded tapes.

Booking Facilities

1. *Crew and equipment.* The PA might be required to book the crew and equipment for the production. In a large company, the PA would book these facilities through an appropriate department, filling in booking forms as required. In a small company the PA might well be negotiating directly with the people concerned. If this is the case, it is important to find out from the director precisely what is required. For example, do you need extra electricians on certain days? If the director does not know exactly what is needed, try to discuss the technical aspects of the booking with someone who is an expert in that area. A lot of money and time can be saved in this way.

When booking crews always make the conditions clear. For example, is it an 'all-in' deal, or will overtime be paid and if so, at what rate? Is tape or film stock included in the deal, or should you purchase it separately? If so, how much should you buy? All these questions must be settled at the outset. You should agree the working hours and travel times and always confirm a booking with a letter giving the precise nature of the booking and the agreed fee, as well as the dates and times required.

2. *Post Production.* You might also be required to book the post production editing and graphics. It is essential that you take expert advice on this as it can be a minefield of different videotape formats, off-lining on linear or non-linear systems, etc. You might find it useful to read the section on Post Production at the back of this book for some initial basic guidance.

Crew and equipment
If the director doesn't know exactly what is needed, try to discuss technical aspects of the booking with someone who is an expert in that area.

Costume and Make-up

While things are daily getting busier in the production office, others are working equally hard in their own fields.

Costume

The first thing the costume designer does is to read the scripts and make notes on the period and the type of costumes needed. If it is to be a period production, much research will follow to ensure historical accuracy down to the smallest detail.

Consultations follow with the director over the designer's ideas, and agreement reached.

The designer notes the continuity of dress from one scene to another in the script and a careful watch is kept on this during the shooting.

As soon as the artists are booked, they are contacted, and measurements taken and discussions held with them. They often have to attend a number of fittings as the work of making costumes progresses.

Moving all the costumes around is a sizeable operation. Often location work can mean cramped, inadequate room for storage, washing, ironing and changing.

As with most things, speed is essential, and it is quite a feat to have a couple of coachloads of extras changed and on the set in double quick time.

Make-up

The make-up artist also reads through the script carefully to determine the period and type of make-up needed. Research is done and notes made on any special make-up, like beards, wigs etc. and any peculiarities relating to continuity, for example, a wound that gradually heals over a number of scenes.

The artists are contacted in the same way as for costume.

During shooting the make-up assistants constantly hover round the set ready to touch up make-up that is wilting under the lights.

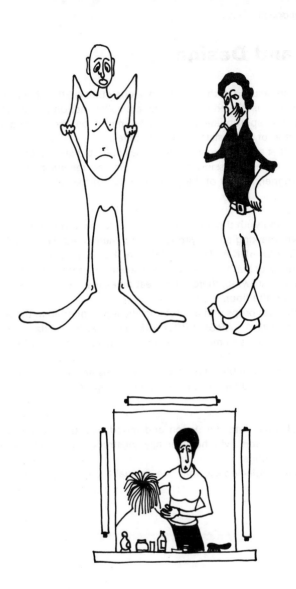

Costume and make-up
Costume have their own problems contending with unusual shaped garments and unusual shaped actors, while Make-up orders wigs, and researches into the correct make-up for the period.

47

Props and Design

Props that are to be used in the action, i.e. a chair that is lifted in a threatening way or a book that is picked up and read, are called 'action props'. These are organised by the appropriate person, who might be the second or third assistant director, or the assistant floor manager.

Other props that are not directly used in the action but are there merely as set decoration are called 'dressing props' and are the responsbility of the art director or designer, as part of the general design of the set.

Action props

The person dealing with them will have gone carefully through the script and the breakdown, marking all the props that are mentioned, i.e. 'JOHN PICKS UP THE BOTTLE OF INK AND THROWS IT ACROSS THE ROOM'. He works out the continuity of such props from one scene to another, i.e. in the scene following the ink throwing, there is no great time lapse, and there should be a large ink stain on the floor.

He also works out props that are not specifically mentioned in the script, i.e. 'MARY IS MAKING A LARGE APPLE PIE'. For that he needs to get all the utensils and ingredients to make the action possible. There must be sufficient supplies to cover any retakes.

He works closely with the art director as the props needed in the action must correspond in style and type to the general dressing of the set.

Design

Much research goes into the period and mood of a production, especially when shooting interiors, whether specially constructed at a studio, or adapted from existing buildings.

Exteriors are simpler unless they are specially constructed streets, house fronts and so on.

Design and props

This set shows the integral part design and props play in the director's overall concept of the production. A great deal of creative effort and research has gone into the design and construction of it.

Shooting Schedule: 1

The focal point of setting up for location shooting is the preparation and assistance in compiling the schedule. This schedule is the consummation of all the preparatory work.

It will comprise all the different pieces of information gathered over the weeks regarding every single aspect of the shooting as well as the day-to-day schedule.

It is a very important compilation, which is sent to everyone associated in any way with the shooting.

Front page
The front page should give the details of the production, the name, the job number, the dates, and also the names and phone numbers of all the unit and everyone connected with the production in a technical capacity.

Cast list
Give an accurate cast list, with the names of all the artists and extras. You might also like to give their agents' telephone numbers on this page. Do not give the private phone numbers of the artists, as the schedule will be widely distributed and the artists might be undesirably harassed if the numbers were too generally known.

Travel
Give detailed information on the travel arrangements that have been made including the times and places transport departs for the location, and lists of the people to travel at that time.

If there is a train service, copy out a train timetable from the principal towns to the location.

For those using their own transport and for guidance to the crews it might prove useful to include maps with the schedule, or at least explicit instructions on how to get there.

Accommodation
Give a full accommodation list, with the hotels' addresses, telephone numbers and prices of the rooms.

Location and contacts
List all the locations, their addresses and people to contact in connection with them. Also some useful addresses and phone numbers would include the local hospital, doctor, hire car or taxi service and police station.

SPIDERS IN THE BATH

by

John Graveney

DIRECTED by LORNA SHELTON

Producer	JONATHAN WHITEHEAD
Production Secretary	PAMELA LOW
Story Editor	JANE GLOVER
Production Associate	ANDREW McGOUGH
Finance Assistant	LARRY HUGHES
Location Manager	ELIZABETH DEAN
1st Assistant	JAMES DELTON
2nd Assistant	BILL SMITH
3rd Assistant	ANGELA DOGHERTY
P.A.	CHRISTINE TURNHAM
Set Designer	LESLIE JOHNSON
Costume Designer	PETER HUNTLY-SMYTH
Costume Assistant	MAVIS WALPOLE
Dressers	SALLY DOWNER/TED WILKINS
Make-Up Artist	SUZIE COX
Make-Up Assistant	JANE LOW/MARY JENKINS
Production Buyer	TED BARCLAY
Lighting Cameraman	FREDERICK SELBY
Camera Operator	JACK WILLIAMS
Camera Assistant	KEITH SHINER
Grips	GEOFF MORGAN
Gaffer Sparks	LES WOLFE
Sound Recordist	GRAHAM ASH
Sound Assistant	HUGH BARBER
Film Editor	JUDY JOHNSON
Assistant Editor	SALLY BROWN
Dubbing Editor	JOHN HARLESDEN

The front page of a shooting schedule.

Shooting Schedule: 2

The way you set out the details of what scenes are to be shot each day is very much a matter of individual preference—yours and the production manager's.

A day to a page
It is sometimes thought simpler to type each day's details on a separate page, which can then be discarded when it is completed.

A week to a page
Others believe that the least amount of paperwork the better, and that it is far clearer to give all the information on as few pages as possible.

Whatever format you favour, there are certain things you must write down:

Rendezvous
Clear instructions on the rendezvous time and place. I have stressed in earlier sections how important this is, as most problems can be sorted out, provided everyone meets up at the beginning.

Shooting order
The shooting order will have been worked out on the basis of the speed of shooting, availability of artists and locations, making the best possible use of the time available.

This shooting order commences with the date on which you are shooting, the scene or scenes you are shooting and then all the relevant details pertaining to those scenes, i.e.: the location, the characters involved in the scene, the time of day of the scene, and whether it is an interior or an exterior. You might also like to put details of the action props to be used.

Rescheduling
While the schedule should be as full a document as possible, changes may become necessary some time during the course of the shooting. You might be held up by the weather, or unforeseen circumstances may force you to make alterations. You might even reach the stage where a complete re-scheduling is unavoidable. If that happens, do check up that any necessary alterations to hotel bookings, transport etc. are made.

The really important thing about the setting-up period is that you should plan for possible contingencies, but if the 'impossible' happens, you should be flexible enough to be able to change things drastically if need be.

<u>DAY BY DAY SCHEDULE</u>   'SPIDERS IN THE BATH'

<u>MONDAY 25 April 1994</u>

0830   MINI-BUS DEPARTS HOTEL, PORTHCAWL (with Gwen and Peter)

0900   UNIT CALL ON LOCATION

1100   MINI-BUS DEPARTS HOTEL, PORTHCAWL (with Maureen,
          Linette, Sarah, Richard and Jonathan)

1230   LUNCH

1900   WRAP

<u>LOCATION</u> : 34 Church Road, Pontypridd, Glamorgan

<u>CONTACT</u> : Mr P. Jones (address above, no telephone)

EP/SC.	PAGE	LOCATION	CHARACTER	SYNOPSIS
2.12	21	Ext. Morn. Garden	JOAN/ALAN	Joan and Alan decide to part
1.35	52	" Aft "	JOAN/ALAN	Joan and Alan first meeting
1.44	63	Int. Night. Lounge	JOAN/ALAN MARY/TED BILL/SALLY JOANNE	Dinner party where Ted accuses Alan of sharp business practice
1.46	69	"	JOAN/ALAN JOANNE	Aftermath of dinner party
2.17	38	Int. Night.	JOAN/ JOANNE	Joanne comforts Joan

<u>TECHNICAL REQUIREMENTS</u>
Black drapes
2nd camera arriving at 12.30

<u>ARTISTS MOVEMENTS</u>
Sandra Grant/Jeremy Sutton arriving hotel in evening

Page of shooting schedule

You are not always in a drama situation.

Documentary-type Productions

The work in setting up an unscripted, documentary-type production follows much the same pattern as for something scripted. Locations have to be found, crews and artists booked and travel, accommodation, catering, etc. arranged. There is, however, an additional aspect to setting up which might involve you to a greater or lesser degree.

Research

Research is the basis of preparation for much documentary filming and your taste for detective work will determine whether you enjoy being involved in documentaries.

The extent to which you are involved in the research depends on who you work for. If you are condemned to 'minding the phone' and typing the letters and given no scope in the preliminary research, you will understandably be bored and frustrated—unless you like phone-minding and letter typing as a full time occupation.

Learning about new things

The aspect I particularly like about the work is the amount you learn about subjects of which you were formerly ignorant. You might well have been ignorant of the subject as it never interested you. You might indeed view with extreme scepticism, as I did, the prospect of a few months research into the history of a famous football stadium, if, like me, you have always kept well clear of organised sport.

But by the end of the research period I was ready to admit that I had learned a good deal that was fascinating about the subject.

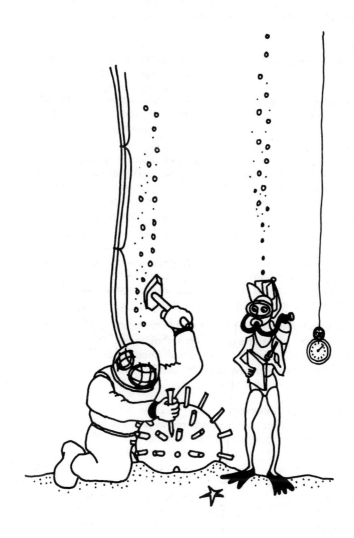

Documentary shooting
Research into documentaries can give you an insight into subjects you might never otherwise explore.

'Find Out All You Can About . . .'

You have started work on a documentary about '....'. You have been given the above directive. Where do you begin?

The subject
Obviously the subject of the documentary will give you the first lead.

Example 1: The history of aviation
A production about the history of aviation could take you on a lengthy journey through libraries, film and video libraries, museums, the air forces of various countries, the historic societies devoted to preserving old planes, etc.

Example 2: The sun-worshippers of S.E. London
A production about the sun-worshippers of South East London will take you to South East London (probably at dawn) to witness the worshippers in action—to a second floor flat in Camberwell which houses the headquarters of the sect—to the libraries for some historical background and to the local newspaper who first brought the subject to light.

Example 3: The training of nurses
A production on the training of nurses would take you to various teaching hospitals, to the library for some background to nursing manuals, and to much discussion with teachers and trainee nurses.

Once you know the subject, you can start researching into the most obvious aspects and other paths of exploration will open up.

The more you dig around, the more you uncover, and the amount of time, energy and patience you devote to the subject depends on the needs of the production and your instinct as a bloodhound.

Other bodies for research
Below is a short list of places which might prove helpful in beginning your research:

Libraries
Town halls
Local councils
Government departments
Law courts
Welfare institutions
Army/Navy/Air Forces
Universities
Polytechnics
Newspapers (national and local)
Film and video libraries

Research
Libraries can be useful starting points for research.

Start packing.

What to Take on Location . . .

As you look round your office with its well-stocked cupboards and shelves, it is very difficult to envisage exactly what you need to take with you on location, and the quantities required. Of course, very much depends on where you are going and for how long, but a selection of the following might be useful:

Scripts and schedules
Always take plenty of these as people are apt to leave them in the most extraordinary places.

Anything to do with the production
Any files specifically related to the production, e.g.: your production file, address book and diary must be taken.

Stationery and forms
If you use any standard forms, for example to do with artists' hours, etc., take enough of them, and a small amount of general stationery—notepaper, envelopes and stamps.

Miscellaneous
Here is a list of easily forgotten odds and ends which you would regret leaving behind.

Typewriter, or laptop PC and printer, clipboard, hole puncher, stapler and staples, adhesive tape, paper and bulldog clips, pencils/ball pens, pencil sharpener, eraser, ruler, rubber bands, spare files, spare folders, rough paper, string, scissors.

In addition for continuity
If your job is combined with that of continuity, take in addition:

Continuity report forms
1 set of coverage scripts
1 set of rough continuity scripts
1 set of scripts (to be interleaved with continuity notes for your daily use)
Progress report forms
Stopwatch
Polaroid camera and film
Daily continuity log forms
A plastic sheet to keep your notes dry in the rain
A small folding stool.

Packing up your office
Do not take too much away with you. You only have the problem of transporting it around from location to location and then probably taking most of it home unused at the end. Work out a list of essentials beforehand and stick to it.

... and Don't Forget the Boots ...

Having packed your mini office to your satisfaction, what about your own belongings? Obviously, I cannot give any specific lists, but below are a few suggestions:

Warm or tropical countries

Suitable clothes, remembering that it is easy to give offence when shooting in countries where dress is related to religious or cultural practice.
Malaria tablets, or whatever preventative medication is advised.
Insect repellent.
A hat with a wide brim.
Sunglasses
Protective sun cream.

Cold countries (including the UK at most times of the year)

Warm clothes, thick sweaters, a windbreaker, thermals, a warm hat and scarf and ... a pair of stout boots.

Don't forget that while most of the unit can go and warm themselves from time to time, if shooting is taking place, you are stuck there on the windy ridge or in the damp forest or wherever, and after a couple of hours you will be frozen.

If you have ever secretly harboured the idea that location work is in any way 'glamorous', you can forget it. It has all kinds of attractions but there can be nothing less appealing than, say, waddling out to a night shoot dressed in every available garment and hardly able to walk with the weight.

Dress for the job
When packing to go away filming, make sure you take suitable clothes for the
conditions you are likely to encounter. Be prepared for hours of standing in the
coldest and wettest places.

Specialised Observation

Having set up the production, you turn up on location on the first day of shooting and do ... what? What is the first requisite for doing the job of continuity?

Before writing the continuity report you must observe what is happening in each shot. So the paramount requirement for the job is having the ability to observe, to notice things. But it is wrong to say 'observation' and leave it at that. One is observing things all the time—certain things make an impression and are retained *in some form or another* in the brain, while other things are seen, but in a generalised way, and no clear imprint is made.

Accurate observation

No two people retain the identical impression of a scene or event. Eye-witnesses at a road accident, for example, are notorious for the diversity of their reports of the same collision. But you can train yourself, or be trained, to observe specific things accurately. Some women workers on egg inspecting lines do nothing but watch eggs pass in front of a strong light, and they are trained to spot defects. Policemen are trained to give fast, accurate descriptions of people. They note certain things; height, build, colouring, clean-shaven or bearded, shape of head, clothes and so on. Likewise with continuity; there are specific things to observe and after a while it becomes second nature to note them.

Much of what you see throughout the day makes only the haziest imprint on the brain. You go into an unfamiliar room. Your eye is caught by a painting on the wall and you remember it afterwards with great clarity, while only having the vaguest impression of the rest of the surroundings.

Training in observation

When looking at continuity in single-camera shooting you train yourself to notice not the random picture your eye happens to light upon but what is important for you to notice and record. Continuity is *not* an abstract exercise in getting it right. It is not a 'spot the difference in these two pictures' game. It is *what* you notice that is important. You will never notice everything that is happening within a shot and it is not necessary that you should, providing the things you *do* notice and write down are those which are important in order to preserve continuity shot by shot and scene by scene.

Spot the difference

No two people retain an identical impression of a scene or event. Note the differences in the above pictures, which represent two eye-witness accounts of the same accident.

Continuity: a Team Effort?

There is the feeling, certainly within television, that continuity is very much a team effort. The costume designer, for example, would therefore be responsible for costume continuity, the props man for action props and so on. Two dangers are inherent in this idea:

Firstly, it is confusing the planning of a production with the shooting. For example, during the planning the costume designer has noted that Scenes 1–4 had direct continuity, and decides that the old tramp appearing in each scene should wear a coat, tied at the waist with string, and a battered hat. When you come to shoot Scene 2, the first in shooting order, the director decides that the hat and coat should be left behind on a park bench. When Scenes 3 and 4 are therefore shot, later in the shooting schedule, the tramp must be without his coat and hat.

But supposing the costume designer was not able to attend the shooting of Scene 2? There might have been a costume fitting to supervise or another urgent appointment. It is also possible that there was not room for all the crew to be present at the shooting due to lack of space.

This is not to denigrate the planning done by the different departments in any way. A great deal of help and assistance are given to continuity by the different members of the unit. But your concern is not with what was planned but what was *actually* shot, scene by scene and day by day.

The second danger is that everyone sees things differently from their own particular interest in the production. Costume will notice costume details to the exclusion of all else, Props will notice props, the cameraman will note the composition and framing of the shot, etc. You are the only person whose responsibility is not to any specific aspect of the shot but to the overall picture, and your concern is how that shot *in its entirety* relates to others *in story order*. You are objective, standing in, in a sense, for the editor and the viewer. You look at all the ingredients that make up the composition of a shot and decide which should have priority over others.

'Costume will notice details to the exclusion of all else'.

Your Role Within the Unit

Avoid being sidetracked

Never be sidetracked into doing anything else when shooting is in progress. Your job is to be where the action is, literally. Continuity implicitly requires you to be there and give your constant attention *all* the time. This does not apply in the feature film industry where the job of continuity script supervisor is clearly defined, but in television you must be clear about your role, as so many other aspects are involved.

Never leave the set and always try to be in a position where you can clearly see all the action. This is usually possible even if it means crouching under the camera.

Continuity is your responsibility

You will find during your first day that everyone knows about continuity. The camera operator says the glass was filled to the brim and positioned on the table. The second assistant insists that the glass was only half full and positioned on the other side of the table. The designer asserts that he left the glass empty on the sideboard and the first assistant swears that there was no glass at all. You *know* that the glass was actually full and in the actor's hand at the relevant moment. (The actor, by the way, remains uncommitted.)

That kind of situation must be stopped before long and involved discussions take place. That is not to imply in any way that no one other than you can be correct about points of continuity. But you are in the unique position of having nothing to do other than watch the action, and if you cannot be more accurate than, say, the second assistant, who might have spent the greater part of the shot stuck behind a bush waiting to give an artist his cue, then you really should not be doing the job.

So do not be perturbed when well-meaning members of the crew come up to you and say, 'Didn't you notice that Fanny's wig was askew during that take?' so long as you *know* that a close-up of Fanny's left foot was being shot.

Always watch the action
Continuity is your responsibility.
Don't ever be sidetracked when
shooting is taking place.

Always try to be in a position
where you can see all the action
even if it means crouching under the camera.

You take the blame.

Making Mistakes

You will of course make mistakes. But a large part of the job is that of having confidence in your accuracy, because then you can get the confidence of the unit and time wasting arguments do not arise.

What to do over a mistake
If you or an artist makes a mistake, and you are immediately aware of it, you can always quietly tell the director and ask for a re-take. Sometimes a director will not re-take for continuity. In that case there is little you can do, except take the rather drastic and ill-advised action of walking into shot, smiling broadly and saying 'Hello Ma'. But after such action you might well find yourself off the set and possibly off the shoot.

If you make an error in a scene already shot and the continuity will affect another scene there are numerous ways of getting round the problem of matching the two scenes. The dialogue could be slightly altered, the first shot for the next scene changed, or even, as happened on the first film I worked on, a new scene written in to overcome a costume error.

Be positive
Always be positive when asked a question. Never fumble through your papers before saying hesitantly: 'Well, I *think* it was with the left hand . . . but of course it might have been the right. . . .' That kind of reply is more irritating than helpful. If you really do not know the answer, say so, positively.

Re-takes for continuity

Except for walking into shot and ruining the take there is little you can do if a director will not re-take for continuity. In such a situation all you *can* do is point out the necessity for a re-take and then keep quiet.

Knowledge of the Script

If you have a script it is very important to know it thoroughly before shooting. You should know, for example, where the scene you are about to shoot occurs in the story and whether it has continuity with scenes before or after. Sometimes scenes occur which do not relate to others. Then your job is easier.

Marking-up script

Read through the script for continuity, marking important junctions before shooting, so that you will be prepared in advance. Stage directions, entering and exiting, the substitution of stunt artists in place of the characters and vice versa are all worth noting, as are points in the script which are obvious indications for a change of angle.

Take this piece of dialogue, for example: '*You think this pen is an ordinary one. But, when I remove the top, and unscrew the base, you will appreciate that it is an ingenious weapon.*' I would use a highlighter pen to mark this dialogue, as the director is quite likely to use a cutaway close-up of the pen at that point.

A-Z

When marking-up the script, it is very useful to go through the scene you are about to shoot and mark each character's dialogue with a different letter of the alphabet (if you get from A-Z and the scene has not ended, then begin again using AA, AB, etc.). Provided you mark up the script for the editor in the same way, this enables you to ntoe very simply where any particular shot begins and ends—rather than having to laboriously write the 'in' and 'out' cues. It also means that you can pinpoint specific lines of dialogue very easily, for example: 'Plane noise F-H'.

Keep up to date

By doing a little homework each night, while shooting is taking place, you can keep comfortably up to date with continuity. For example, Scene 4 has direct continuity with Scene 5 in that, let us say, two characters are shown in both scenes and there is no time gap. Today you completed shooting Scene 4. Scene 5 will not be shot for another fortnight. If, at the end of today, you sit down with the script and mark in pencil any continuity notes at the top of Scene 5, i.e. '*John and Jill must enter frame right. John wearing blue sweater, green jeans and carrying suitcase in left hand. Jill wearing red/white jacket, blue jeans, red/blue striped blouse (two buttons open at neck) and carrying shoulder bag right shoulder*', then, in two weeks' time, when you come to shoot Scene 5, you will have all the information at your fingertips and not have to scrabble about in the script in order to find the answers to the questions asked by the director.

Script interleaved with notes

The system I use is to carry the script with me, in story order, with the pages of continuity notes of the scenes already shot interleaved with the script pages. Likewise, I would keep any photographs of costume or props together with the script pages for that scene. That makes it easy to check on any continuity queries I might have, as all the relevant notes are together in one file.

70

SCENE 505 : EXT. STREET
PAULINE WALKING

SCENE 506 : DAY. INT. OFFICE

LAURA IS SITTING BEHIND HER DESK, WHICH
IS LITTERED WITH PAPER AND
PHOTOGRAPHS. EVEN THE CHAIR IS FULL
OF PAPERS. THERE IS A KNOCK ON THE
DOOR.

(A) **LAURA**
Come in!

PAULINE ENTERS THE ROOM AND
APPROACHES THE DESK

(B) **PAULINE**
I've come about the advertisement.

(C) **LAURA**
Just a minute. (SHE FINISHES WHAT SHE IS
DOING BEFORE LOOKING UP) You've come
about...?

(D) **PAULINE**
The advertisement. You know, the one in the
paper.

(E) **LAURA**
Oh, of course! (SHE SHAKES HANDS) Do sit
down. Just put that junk anywhere. I always work
in an awful mess, don't you? Now you're...?

Work out a fast reference system

Work out your own system to enable you to check continuity points quickly from
scene to scene.

When shooting the second scene (shown above as the second picture) you must be
able to check the important points quickly—how the lady is dressed and which way
she should enter frame.

Coverage Planned

After getting to know the script, try to find out the director's plans for coverage of a scene.

Important junctions
If you know where the important junctions for cuts will be, you can pay particular attention to the continuity at those points.

The director, of course, may not have any plans, except, one assumes, in his head.

Circulated shot lists
Almost every director I have worked with has started shooting in an incredibly efficient way by having lists of their proposed shots circulated to the unit. After a few days the unit thank you profusely for these lists, fold them up neatly and put them away 'in a safe place', never to be looked at again, and the director then confines his lists to himself and you.

After two weeks or so, even these lists, which have become vaguer day by day, cease, and you must find other ways of determining what sort of shot is being planned.

Shooting script
There are occasions, however, when you might find yourself working for a very well organised director, one who plans well in advance exactly what he is going to shoot and works out each angle with meticulous care. You should of course, think yourself very lucky to be working for such a paragon—but do watch out for one thing.

Such a script can be of immense help in giving a very good idea of what the director *intends* to do, but do beware of following it too slavishly as it may well differ quite considerably from what he *actually* does on the day of shooting. Your job is to note down what actually happened and not what your director's intentions were some five weeks or even five days before the shooting.

Storyboard
Some directors work out a storyboard—sometimes with the help of the graphics designer. A storyboard plots out the story in picture form, showing all the different angles the director plans to shoot. It is a very useful tool as everyone can instantly understand a picture, even if it is just a very simple line drawing, whereas some people might find it harder to visualise a written shot description. The storyboard is used extensively in planning commercials.

The director's plans
There are a number of ways of finding out how the director plans to cover the script.

How to Keep Informed

Look at the shot itself
When shooting on videotape, there is likely to be a monitor on location. This is immensely useful as it makes the job of giving an accurate shot description very simple. There is also the facility, in matters of dispute over continuity, of being able to play back the tape. But this facility should be used sparingly, as it wastes time on location and can lead to the general cry of 'let's play it back' over the smallest query.

When shooting on film, the camera may be fitted with a video-assist system in which case there will be a monitor for you to look at. While continuity can be done quite efficiently from a monitor, it is important to be able to work without one, otherwise you will experience enormous difficulties if there is no monitor on your shoot.

If you do not have access to a monitor, you can keep informed as follows:

Stick by the director
At some point the director must communicate with the cameraman, so if you have no idea what shot is to be taken, keep close to the director and cameraman and you will find out.

Master shot then cutaways
Frequently the first shot that is taken in a scene will be some kind of master or establishing shot and the rest of the coverage will follow. But there are no hard-and-fast rules determining how a director will shoot any particular scene.

Position of the zoom handle
When shooting on film, the camera operator may be using a zoom lens by which he can adjust the magnification of the picture or, in other words, the amount of the scene that is included in the shot. Provided the lens is manually operated by a handle, you can judge the size of the shot by the position of the handle. But do find out first whether the wide-angle setting is at the top or bottom of the lens, as they vary. Video cameras have a motorised zoom.

Keep close behind the camera
Always stand as close to the camera as possible. You cannot describe a shot accurately if you are looking at the scene from a different angle from the camera. Do not set yourself up with a folding table and camp stool some way away from the shooting. You may be comfortable but your notes will be wildly inaccurate.

Look through the viewfinder
No cameraman should object if you ask to look through the camera viewfinder during a rehearsal, provided you choose an opportune moment.

Always ask
Finally, if all else fails, you can always ask the director or the cameraman.

Keep close to the camera
By keeping as near the director and cameraman as possible you not only keep abreast
of what angles are being discussed and taken, but are also in a position to describe
the shots more accurately.

How to Describe Shots

An accurate shot description is the first essential of continuity. Always describe the shot from the camera's viewpoint (which is also your viewpoint and that of the audience). For example, *'Fred enters frame camera left and exits bottom of frame right'*.

The following gives a list of the most commonly used shot descriptions and abbreviations:

W/A Wide-angle shot. Such a shot takes place in a wide area of the set in front of the camera. It is sometimes referred to as a VLS (very long shot).

LS Long shot. A shot which directs the viewer's eye to the depth rather than the width of the shot.

MLS Medium long shot. Refers to a shot comprising the head to just below the knee of the subject.

3-s Three shot. A shot containing three central characters.

2-s Two shot. A shot containing two central characters.

2-s fav. X A shot with two people—the camera favours one person more than the other.

o/s 2-s Over the shoulder two shot. Two people are seen in the shot but the camera is looking at one of them over the shoulder of the other.

Mid 2-s Comprising the head to just below the waist of two people.

Close 2-s Comprising the head and shoulders of two people.

Deep 2-s A shot containing two people—one in the foreground and one in the background.

MS Mid shot. A scene at normal viewing distance. In the case of a human subject the camera frame cuts the figure just below the waist.

MCU Medium close-up. The camera frame cuts the figures at chest level.

CU Close-up. The camera frame cuts the subject just below the neck.

BCU Big close-up. The face fills the frame.

X's POV X's point-of-view shot. The camera is X and sees as if from his point of view.

H/A High angle. The camera is above the action and looking down on it.

L/A Low angle. The camera is below the action and looking up.

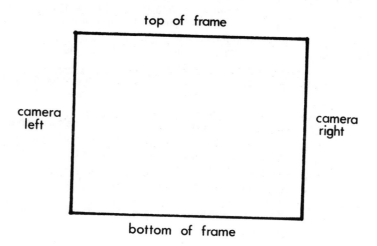

Exits and entrances

A point to remember is that when referring to entrances and exits always do so from your (also the camera and audience) viewpoint.

Other Terms Used

Camera Movements

PANNING	Camera turns from one side to the other, pivoting horizontally on an axis, either right to left or left to right.
TILTING	Camera pivoting vertically on an axis, tilting up or down.
TRACKING	Camera is physically moved forward or back, towards or away from the subject.
CRABBING	The camera is physically moved crab-wise or sideways to the direction of view.
CRANING AND JIBBING	A movement by a camera mounted on a crane dolly. The dolly has a jib arm which can be raised and lowered rotating around its fulcrum.
Z/I	Zoom in. The camera is not moved but the focal length of the zoom lens is increased. This magnifies the subject without changing the perspective of the scene (as opposed to a track where the camera moves towards the subject and the perspective changes as if you were walking towards it).
Z/O	Zoom out. The lens is adjusted in the reverse direction from the above.

Abbreviations relating to action

A/B	As before
FAV	Favouring
F/G	Foreground
B/G	Background
F/WD	Forward
B/WD	Backward
X's	Crosses or across
CAM R	Camera right, i.e. as seen from the camera's—and the viewer's—position when facing the action
CAM L	Camera left
O/S	Over the shoulder
OOV	Out of vision
OOFL(R)	Out of frame (left or right)

Abbreviations relating to sound

MUTE	Without sound
SYNC	Sound recorded synchronously with picture
W/T	Wildtrack (sound recorded without picture)
V/O	Voice-over
MIC	Microphone
F/X	Effects
ATMOS	Atmosphere

Description of shots

1. Wide angle (W/A). 2. Long shot (LS). 3. Three shot (3-s). 4. Two shot (2-s). 5. Two shot favouring X (2-s fav. X). 6. Over the shoulder two shot favouring Y (o/s 2-s fav. Y). 7. Mid shot (MS). 8. Medium close-up (MCU). 9. Close-up (CU). 10. Big close-up (BCU). 11. High angle (H/A). 12. Low angle (L/A). 13. Tracking (the camera is mounted on a dolly and physically moved). 14. Panning (the camera turns from one side to the other). 15. Tilting (the camera moves up and down).

Other Terms in General Use

C/A (Cutaway)
This can apply to any shot that is not a master; for example, a close up. Such a shot often emphasises or highlights a particular aspect of events when it is inserted in the main action.

Cutaways are also used as a safety measure. For example, a woman is telling a long story. Several cutaways are taken of her audience. The woman's story can then either be shortened by cutting away to the audience, or the cutaways can be used for dramatic effect—to build up atmosphere, heighten tension and generally give pace to a scene.

Under or over-cranking
On TV, film is shot and replayed at a standard rate of 25 frames per second. Feature films are projected at 24 frames per second. That is to say that any departure from this standard speed of shooting alters the speed of the action. Under-cranking is a term to denote shooting at below normal speed, to speed up the action on replay. Conversely, over-cranking at a higher speed than normal, slows down the action on replay. The 'cranking' part of the term originates from the days when cameras were cranked by hand.

When shooting on video, unless extremely specialised equipment is in use, the subject is recorded at normal speed and the videotape replayed at a different speed at the on-line edit. For example, a car which can only travel at 30 mph needs to move at 60 mph in the finished product. On film it would be under-cranked at 12 fps, i.e. roughly half the normal rate, so that when the film is replayed at 25 fps it will appear to be moving at just over 60 mph. Using videotape, the car will be recorded normally and the tape replayed into the final edit at a speed of + 200%, again giving the car the appearance of travelling at 60 mph.

Process shots
In both film and video productions shots are often taken on location which will only form part of finished images in the final production. Simple examples include a point of view along a road which will eventually appear through a car windscreen and a group of people sitting on sunbeds in a studio on the right-hand side of frame which will eventually be blended with a location shot of a hotel swimming pool.

1   2

3   4

Other terms used
1. Cutaway from W/A of guilty man looking towards bookcase. 2. Cutaway from a close up of a woman screaming. 3. A woman is talking to an audience . . . 4. . . . a cutaway is taken of the audience to use: (a) to shorten the woman's chat. (b) for dramatic effect.

Know What to Observe: 1

Good continuity is not *just* being good at observation. It is knowing what is important to observe.

Once you know what the shot is you can observe as necessary.

Close shots

It really is rather a wasted exercise giving detailed information about the clothes an actor is wearing when a large close-up of the face is being shot. It is also unnecessary to worry too much about the position of the props in such a close-up, unless the props are in frame or brought into frame, i.e. a close-up of someone drinking or a cigarette being brought up to the lips.

Mid shots

Be aware of *exactly* what is in shot. For example, someone is reading a book and turning the pages over frequently. In a close-up, all you see is a face, frowning with concentration; in a wide shot you see the whole action with the book in shot. In a mid shot, even though the book itself is not visible do not ignore the operation of turning over the pages, because the upper arm and shoulder movements will be noticeable.

Wide shots

In a very wide shot, it isn't necessary to be as observant of the minute details of actions and props as in the closer shots. In a wide shot of a pub interior with many actors sitting and standing with their drinks, the exact levels of the drink are unimportant, as long as something is in their glasses.

Very long shots

In a very long shot it might prove to be immaterial whether or not there is any drink at all, as it is impossible even to see the glass.

Size of shot

A   B

Close shots
Only worry about the clothes or props when these are in frame (A, B).

C   D

Mid shots
It is important to know exactly what is in shot. All you see in the close-up is the face (C). In the wide shot you see all the action (D). In the MCU you will see the upper arm movement as the pages are turned over (E).

E

F

Wide shots
You do not have to be as observant of the minute details in this busy station shot (F).

G   H

Very long shot
In this long shot it is just possible to see the three figures—let alone details of clothing and props (G, H).

Know What to Observe: 2

It is generally true to say that the more that is happening in a shot, the less continuity matters. But in a constrained, intimate situation, every head turn counts.

Intimate situations

Two characters are having a quiet *tête-à-tête* over dinner. They are eating, and sometimes talking, but nothing much is happening visually. In such a situation it is most important to be accurate about all the actions, even head nods and slight gestures, particularly where the director has planned to cover the scene with complementary two shots. In these shots both characters are in vision all the time and it is only possible to cut from one to the other when the dialogue and action match exactly on each shot.

If the director has planned a coverage of single shots on each person, the action is not so critical as most of it is out of view. The props also are of great importance in such a scene.

The opposite extreme to an intimate situation in terms of observing detail for continuity would be a fight sequence, including a large number of people and props. It would be impossible to notice everything, especially head nods, in a situation with such violent activity. But in this kind of set up you should note carefully what the principal actors are doing and try to keep track of their relationship with other groups.

For instance, in the early part of a scene two principals may be fighting with each other in the thick of other extras. In the latter part of the same scene the principles are still fighting it out, but the extras have cleared save for a group on camera right. So remember that.

Also, be aware of clothes becoming disarranged during a fight and possible make-up requirements, like blood and bruising, that may have continuity with other shots and scenes.

84

1. In a scene with little action, every detail of continuity is important.

These two complementary 2-shots will only cut together if continuity matches exactly in every detail.

2. But in a scene with a lot of action, accuracy over every detail is not so vital.

Just pay attention to the principal characters and have a general ideal what else is happening in the shot.

Priorities

As you cannot notice *everything* that happens in a shot, what should you concentrate on? Below are a few guidelines:

Largest moving object
The largest moving object within the framework of a shot is the most important to observe as the viewer's attention will be drawn to it—whether it is a head seen in close-up, a person seen in mid shot or a large pink elephant entering frame and dominating the wide shot.

The main characters
If there is a great deal happening in the shot then you should concentrate on the main characters. Stick to watching the people the production is about, who will hold the audience's interest and attention. Others in the shot only assume importance *in relation to* the main characters.

The person speaking
Let us suppose you are doing continuity on a scene containing six principal actors. They are sitting round a table, talking, eating and drinking. How can you possibly notice what they are all doing at any one time? The answer is quite simply that you cannot. But if you ensure that each time the scene is repeated all the props are reset correctly, then the actors will repeat their actions at more or less the same time. Another point to note is that an actor is unlikely to take a bite of a ham sandwich before launching into a speech. But most importantly, you should watch the actor who is talking because the shot used will be more likely to favour the one who is speaking. That is, of course, until the large pink elephant walks into shot!

Bright colours
Bright colours attract the eye, especially if surrounded by drab ones. So notice any areas of bright colour in the shot.

Shots travel in and up
The most interesting part of anyone is the head—unless you are working on a shoe commercial! Coverage of a scene tends, therefore, to travel from the wide shots to the close-ups—inwards and upwards. Thus it is more important to notice the upper part of the body than the lower part and to observe head and arm movements in preference to props littering a table.

'The largest moving object within framework of a shot is the most important to observe as a viewer's attention will be drawn to it . . .'

Props

Ask anyone not connected with shooting—and a good many people who are—what they immediately think of when the term 'continuity' is mentioned. Many believe it is a question of avoiding an accidental total change of costume from one shot to the next, or props appearing in different places from shot to shot, or things suddenly disappearing ... and so on.

There is a good deal more to continuity than maintaining consistency in costume and props, but those are often the most noticeable.

Dressing props
Dressing props are those used to dress the set but are not handled by the actors. Take a Polaroid picture of the set or draw a diagram. This is useful if you are working for several days on the same set or for reference in the case of a shot being retaken.

Action props
Action props are those used by the artists in the course of the action of a scene. It is most important that you pay particular attention to these props: when the artist uses them, which hand they are held in and at precisely what point in the dialogue these actions occur.

Resetting props
If the action props are reset in exactly the same way then the artists will integrate their use with the dialogue and they will be used in the same way and at the same time in the scene. If the props are reset in different positions, the artists will not get used to handling them and the timing of the actions in relation to the dialogue will be different.

Note the general position of the props by drawing a diagram or taking a Polaroid photograph, but only worry about the props that will be used in the action.

Costume

More problems seem to arise over costume than almost anything else, but maintaining costume continuity should not be difficult providing the following points are adhered to.

Advance planning
The costume designer, in advance of shooting, will have worked out the scenes with continuity of costume in mind. These details will probably appear on the script breakdown in the form of days, i.e. Day 1, Morning; or Day 1, 10.00 am, etc. But during the actual shoot there may be changes of costume which are unplanned. For example, the director might decide that an actor should take off his coat in the middle of one scene. This had not been pre-planned by the costume designer and might well have a knock-on effect with costume continuity in later scenes. This is where your notes and photographs become essential.

Take a polaroid
At the start of shooting a new scene, i.e. one with no prior continuity, take Polaroid pictures of the artists involved.

Write it down
Always write down exactly what the artist is wearing at the start of each fresh scene, and make your notes as full as possible.

'Mrs Bold: coat, hat, dress', does not convey very much. Is the coat done up or undone? Are all the buttons done up? What sort of coat is it? What colour? Likewise with the dress. Is the hat on her head or in her hand? What about her hairstyle? What about shoes? Does she have any accessories, such as handbag, gloves, etc?

Checking
Having taken photographs and made extensive notes, then use this information to check when matching continuity of costume. This is the absolute key to all good continuity. It is not enough just to make notes and take photographs whether of costume, props or sets. Those notes must be used intelligently to match continuity from shot to shot and scene to scene.

My first continuity error was over a scarf that was being worn in one scene and not in the next. My second was over a pair of Wellington boots. No one may notice these mistakes, or they might achieve unwelcome fame. With a little care and trouble you need never make them.

Make a note or take a Polaroid photograph of costume. You can easily mislay a few items of clothing—especially with period costume.

Not only has the soldier lost his hat—he has also lost his bag, his sword, and the top of his bayonet. It's obvious when the pictures are placed together as these are, but you could easily make the same mistake in the real situation and you had not made copious notes or taken a photograph.

Continuity of Action

Some of the least obvious aspects of continuity are no less important than those already discussed. But they are often overlooked because they are not often apparent in the finished production. The reason they are not apparent is usually because the editor has found a way round the mistakes in one way or another, but in doing so, the artistic role of the post production editing process may have been severely limited. Such errors put an unnecessary and time-consuming burden on the editor.

Let us examine these aspects of continuity by starting with the simplest—continuity of action.

Example 1: A girl walks across a bridge, putting her coat on in the process.

The director has planned two camera angles on this action: one from the bank, with the girl walking away from the camera, and the other from the opposite bank, with the girl walking towards the camera. The cut from one camera angle to the other takes place when the girl is somewhere on the bridge.

Suppose in shooting these two angles, there was an error of continuity in the action of putting on the coat. When the girl put on the coat in the second shot, she was in a different position on the bridge from her position in the first shot.

If those shots were then cut and joined together at the logical place, where the director had planned it, the result would show a jump in the action—a jump cut.

It is possible in that situation for the editor to 'cover up' this error of continuity by cutting earlier or later. But that would not give as smooth a cut as if the action had been repeated correctly in the second shot. In other words, the cut had to be made where it was mechanically possible and not in the way that was most suitable dramatically.

Example 2: A man opens a door with a key. He enters and closes the door behind him.

There are two camera angles—one on each side of the door. The second angle is taken much later and as the action with the key has been shot on the first angle, the door is set for the man to just push it open for the second shot. On that angle he pushes the door open and walks in and everyone accepts the take. *But he uses a different hand on the door* and the editor is faced with the same problem as before.

1st angle

B

C

2nd angle
Note that the coat is put on in
a different place in the bridge.
It would not cut together with
the first angle

This would have been the correct
place for putting the coat on

D

1st angle
Man approaches door (outside)
(D).

E

F

2nd angle
From the other side of the door
(inside) (E, F).
Continuity of the hands differs
from the first angle. These shots
would not cut together

The position of the hands are
correct in this shot (G)

G

Continuity of Action and Dialogue

What happens when there is dialogue as well?

Example 1: A presenter of a programme is coming to the end of a long piece spoken 'to camera'. He moves off, out of frame *after* saying 'We pass on to the next exhibit'. The second angle is a wide exit shot, and he repeats the end of his piece and then moves off. But he starts to move off *on* the words 'We pass on . . .'.

If those two shots were cut together there would either be a loss of action (a jump cut), or double dialogue: 'We pass on' on the first angle, and again 'We pass on' repeated on the second angle.

In such a case the editor would either use the ugly jump cut, or stay on one shot only. He obviously cannot repeat the dialogue. Either way it spoils the intended effect and, if only one shot is used, makes the shooting of the second shot totally pointless. One often sees that a jump cut has been used to get over this type of error. But it is quite wrong to force the editor into using such a jarring effect because of sloppy continuity.

To make the relationship of action to dialogue absolutely clear, take this well-known nursery rhyme:

Example 2: A girl is walking along, swinging a shoulder bag and reciting:
'Mary had a little lamb,
Its fleece was white as snow.
And everywhere that Mary went
The lamb was sure to go.'
On the first angle she places the bag on her shoulder *after* saying: 'Its fleece was white as snow'.

But on the second angle she places the bag on her shoulder *before* saying: 'Its fleece was white as snow'.

When those shots are cut together there is either a sudden jump in the action—from the girl with the bag in her hand, to the girl with the bag on her shoulder, or the action is perfect and the dialogue is as follows:

First angle
'Mary had a little lamb
Its fleece was white as snow'

Second angle
'Its fleece was white as snow,
And everyhere that Mary went
The lamb was sure to go'.

Mary had a little lamb

Its fleece was white as snow

And everywhere that Mary went

The lamb was sure to go.

Action and dialogue

On the first angle the girl places the bag on her shoulder *after* saying 'Its fleece was white as snow' and on the second angle she places it on her shoulder *before* saying the same dialogue.

If those shots are cut together there will either be a jump in the action, or double dialogue.

Drama Situations

It is not often, however, that you have only one character to concentrate on. Things get more complicated as more characters are involved.

Example: A man is sitting on a seat. A girl enters frame—she walks up to the man. She says 'Hello'. The man replies likewise. He gets up and they exit.

That could hardly be simpler. But endless variations of that simple scene are possible.

Does the girl stop *before* she speaks, *as* she speaks, or not at all?

Where does she stop?

When does the man get up?

Does he speak from the seat, on the rise, or standing?

What is the relationship of the man and the girl as she speaks/as he speaks/as he rises?

And so on.

If that simple scene were shot from two angles with both characters in shot in each angle, the continuity has to match exactly on each angle.

first angle   second angle

1   1

2   2

3   3

The simple drama scene
On each angle the action in relation to the dialogue must match exactly, otherwise the shots 1, 2, 3 will not cut together.

Crossing the Line

Avoiding 'crossing the line' is one of the conventions of single-camera shooting which comes under the umbrella of continuity. The 'line' is an imaginary one, drawn between the noses of two people as they look at each other. Providing the camera does not cross that line, shots that cut directly onto each other will match.

John and Jane are sitting opposite each other. John is on the left of frame and Jane on the right. These positions have been established in the first shot that was taken. Any other shot cutting directly onto or into that shot must feature John on the left and Jane on the right of frame, otherwise it would appear, when the shots are edited together, that the two people have changed places—in other words, they would have jumped from one side of frame to the other.

Single shots

This convention is applicable even to single shots. An interview is being held with the interviewer standing out of shot, to the right of the camera. The interviewee is therefore looking foreground right (to the interviewer). When the shots of the interviewer are taken, he must look foreground left in order that his shot will match that of the interviewee.

Shots that cut directly onto each other

Because this convention only applies to shots that cut directly onto each other the director should work out the coverage of a scene with care, to ensure that when the scene is edited the shots will cut together without crossing the line.

Exceptions

The 'line' can be crossed in the following situations:

1. When the camera moves during the course of a shot, establishing a fresh relationship to the characters;
2. When the characters move, establishing their own different relationship to the camera;
3. When there is a fixed point of reference for the viewer which is unchanged, i.e. a doorway;
4. When the size of shot changes dramatically from one angle to the next.

Both 3 and 4 are known as a 'direct reverse'.

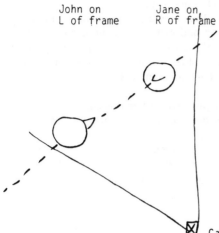

John on
L of frame

Jane on
R of frame

Camera position noted and angle

Dot in a line between people's noses as they look at each other. Providing the camera does not cross that line, shots that cut directly onto each other will match: i.e. John and Jane will remain on the same side of frame as in the original and will not 'jump' in frame.

This direct reverse works. Why? They have their backs to the camera and the bride is on the right. Inside the church the next shot shows them facing the camera. Their positions are reversed.

Screen Direction

If an object or person is to appear to travel from destination A to destination B seen in a series of shots of differing angle then it must continue to travel *in the direction* that has been established in the first shot.

A train is passing through a station. The camera, positioned on Platform 1, sees the train enter frame from the left and exit right. If the camera is then repositioned on Platform 2 it will see the *same train* entering from the *right* and exiting *left*. If these two shots were subsequently cut together, the viewer would think either that the train had changed its mind and changed direction or that it was another train.

Our old friend 'crossing the line' is again involved. In this case the direction of movement is the railway track. If the camera crosses the track then the direction of movement appears reversed.

Because the viewer only sees what the camera shows, *from the camera's position in relation to the subject*, there is no external evidence to indicate that the camera itself has changed position, unless this is made apparent in some other way.

Changing direction

Therefore if one has established movement from left to right across the screen this direction must be continued unless:

1. The object changes direction in front of the camera;
2. The camera takes the viewer to a different position by tracking round the moving object;
3. The object travels straight towards the camera or directly away from it;
4. A cutaway is imposed;
5. A shot is taken from within the moving object, i.e. inside the train or car. But beware in this instance of ensuring that the background is moving in the correct direction!

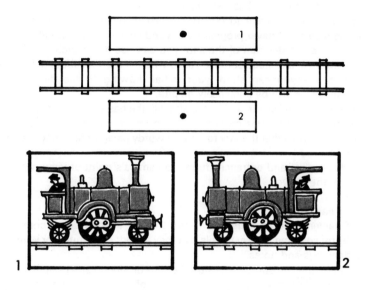

1. The shot of the train taken from Platform 1 and the shot from Platform 2 show the same train apparently travelling in opposite directions.

Drawings and Diagrams

Always draw a quick, rough sketch of each shot. Draw the heads of the characters, denoting their positions in relation to the camera, and their noses, and dot in an imaginary line. Show, in addition, the main props and set dressing. You do not have to be good at art in order to draw a perfectly acceptable diagram. I never progressed beyond stick men at school!

Once you have a diagram of the shot you can use it to match subsequent shots. If you are at any stage confused as to whether the shot about to be taken crosses the line you can show your initial drawing to the director and cameraman, and it will be immensely helpful.

Everyone relates to pictures—that is, after all, the basis of this industry—and it is far easier for you and everyone else to match pictures represented in diagrammatic form rather than try to unravel wordy paragraphs giving the same information.

Note down exits and entrances from the camera's point of view, which is also that of the viewer.

Polaroid cameras

Polaroid cameras are an extremely useful tool for anyone doing continuity, especially for costume and set details, but they are not substitutes for your own simple drawings and notes.

Don't think that by standing beside the film or video camera and taking a picture you will necessarily get a photograph of the same size as that being shot by the cameraman. If you want a close-up photograph of the set dressing on a table go close to the table, possibly stand above it, and then take a picture.

Try to ensure that you have a camera that is in good working order and look after it on location. Don't forget to take plenty of film with you, as it might be hard to obtain in Outer Mongolia!

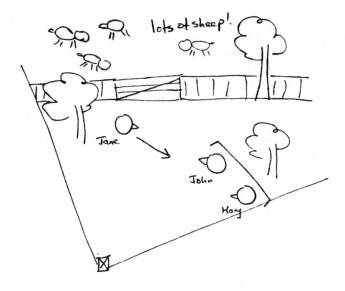

lots of sheep!

Jane

John

Mary

Always do a rough diagram of the shot, showing the camera position as well as the actors and main props. It does not matter how rough your diagram is, it will be of enormous help in matching future shots.

Rehearsals and Actors

Rehearsals are vitally important as they provide you with your only opportunity of writing down all the information you need. You will never be able to do the job properly if you ignore the rehearsals and only pay attention during the actual take.

There is so much you need to notice in any shot that you can only build up a complete picture bit by bit. Rehearsals afford you the opportunity to do this. For example, during the first rehearsal of a scene, you may not be able to write down the action, but you can at least make a note of the actor's costume. You can also take a Polaroid or other instant picture or draw a diagram of the basic set and the approximate camera position. At the next rehearsal you can note down a bit more—perhaps the overall action, and slowly you will find that you have built up more and more information until you can concentrate entirely on the specific actions and props. Things will be bound to change from one rehearsal to another, but it is far simpler to alter your notes accordingly as and when the changes happen than it is to start writing from scratch during a take.

Problems with actors

If an actor is good with props and naturally does the same action at the same time and with the same bit of dialogue you can stop worrying about him, except to remind him, if requested, of movements when the master shot was taken some time back, or in scenes where the action is complicated.

If an actor is not consistent, watch him or her like a hawk. Don't worry during the rehearsals, but write down the exact movements during a take, so that you can go over them later for cutaways—extra shots of him to be inserted in the main action, if needed. Do not let anyone else tell the actor what he did and when. He will only get confused between your advice and everyone else's. You get in first. It's your job.

The difficult actor

Watch out for the 'difficult' actor. He tends to be terribly nervous and unsure of his actions. He will constantly come up to you to ask anxiously what he did with his left hand in the last take, and will probably worry incessantly about everything. Always have an answer ready. Maybe you did not notice what he did with his left hand. Perhaps it was not even in shot—but never let him know that you have not recorded his every move, in or out of vision. He needs reassurance and will stop worrying only when he feels he can rely on you. You need to keep a careful eye on him for another reason as well; he will argue to the end if he thinks you are wrong about something.

Some actors are good with props . . .

. . . and some have to be watched carefully.

Actors and their props

The Complete Scene

We have now looked at the different ingredients to observe for continuity—action, dialogue, props and costume. Now we shall put them all together in the following scene:

Fred and Mabel
Scene 1: Kitchen: day
MABEL IS PREPARING AN APPLE PIE.
FRED COMES IN FROM THE GARDEN AND RUSHES OVER TO THE SINK. HE TURNS ON THE TAP AND HOLDS HIS HAND UNDER THE STREAM OF WATER.
MABEL GOES OVER TO HIM.

MABEL: Fred, Fred, what's happened?

MABEL EXAMINES HIS HURT HAND.

MABEL: How on earth did you do that?

FRED: The bloody adjuster stuck on the lawn mower.

MABEL: You've got dirt in it.

SHE PUTS HIS HAND BACK UNDER THE TAP.

MABEL: I'll get you a plaster.

SHE GOES TO THE CUPBOARD AND GETS OUT A TIN OF DRESSINGS.

MABEL: Do you want some antiseptic on it?

FRED: No, don't bother.

MABEL TAKES A TOWEL AND GIVES IT TO FRED. HE DRIES HIS HAND ON IT.

FRED: Thanks, love.

MABEL STICKS DRESSING ON FRED'S FINGER. HE WINCES.

MABEL: Did I hurt you?

FRED: No. It's just a bit sore. (PAUSE) Right. I'll finish the lawn.

BOTH GO TO THE BACK DOOR.

FRED: What time's lunch?

MABEL: Is half an hour all right?

FRED: Fine. Give me a shout.

HE KISSES HER AND GOES OUT.

Scene 2: Garden: day
FRED COMES OUT OF THE HOUSE AND WALKS TO LAWN MOWER . . .

The scene is set for Fred and Mabel.

Fred and Mabel: Coverage Planned

The director has planned the following shots for Scene 1 and he has further planned to shoot them in this order:

1. Wide angle of whole scene (master)

2. Tight 2-s Fred and Mabel at sink

3. MCU tap and hand

4. CU dressing going on to finger

5. CU Fred wincing

6. MCU cupboard—tin being removed

7. MS Mabel looking pleased with pie

8. 2-s Fred and Mabel at door

Without being able to show the action in the master shot, I shall try to describe it as fully as possible. When actually doing the continuity for the master, you should have written or noticed as much of the following as possible. There are some obvious pitfalls and important things to notice for the other angles and I have mentioned these where necessary.

Number of shots planned

1. Wide angle (W/A). 2. Tight two shot (Tight 2-s). 3. Medium close-up tap and hand (MCU tap and hand). 4. Close-up plaster (CU plaster). 5. Close-up Fred (CU Fred). 6. Medium close-up cupboard (MCU cupboard). 7. Mid-shot Mabel (MS Mabel). 8. Two shot at door (2-s at door).

Wide Angle: Fred and Mabel Scene: 1

Action and sound

Important to observe

MABEL standing behind the kitchen table puts the last thumb print in the pie and stands back to admire her work.

How does she stand back? Wiping hands on apron or what?

She turns to the oven and opens the door.

Which way does she turn?
Note the oven door action. Is the door left open or shut?

She turns back to table and picks up pie.

FRED comes in through back door (which is closed to start), he leaves it open, crosses behind Mabel and dashes to sink. He turns on the tap with left hand and puts finger of right hand under the jet of water.

Back door closed to start—left open by Fred—keep an eye on it.
Which tap does Fred turn on and with which hand?
Which finger of which hand is cut?

MABEL puts pie back on table—closes door of the oven with her knee in passing and hurries over to Fred.

Pie does not figure again, but it is left on the table so do not forget that it is there.

Note oven door action.

MABEL: Fred, Fred, what's happened?

Does she say that line on the move or when she is by Fred's side?
Which side of Fred does she end up on?

FRED doesn't answer. Just swears under his breath.

MABEL is now at the sink. She takes his right hand with her left.

Which hand takes which? Important to note that the tap is left running.

MABEL: How on earth did you do that?

FRED: The bloody adjuster stuck on the lawn mower.

MABEL: You've got dirt in it.

She sticks his hand back under the tap.

During the above dialogue, does she hold his hand all the time? Does she bring his hand up to her face to look closely at the wound, or bend down to look. When *exactly* does she stick his hand back under the tap?

110

MABEL: I'll get you a dressing.

She crosses to the cupboard—opens the door (right hand to open it)—takes out tin of dressings (with right hand)—transfers tin to her left and says:

MABEL: Do you want some antiseptic on it?

FRED: No, don't bother.

Note all her actions at the cupboard—which hand does what. N.B. There is a bottle of antiseptic in the cupboard. Make sure it is still there for the cutaways.

Does she say that line while transferring the tin to left hand or after?

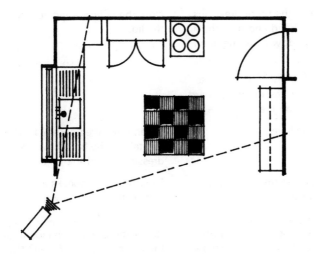

Plan view of kitchen with camera position showing complete scene as per page 106.

Wide Angle: Fred and Mabel Scene: 2

Action and sound	*Important to observe*
MABEL leaves the cupboard door open—moves to the back door—half closes it and with her right hand picks up a towel from a hook on the back.	Note that cupboard door. I said to keep an eye on the back door.
She moves forward to the sink. FRED turns off the tap with his left hand and takes the towel from Mabel also with his left hand.	Tap action. At what point does Fred turn it off? i.e.: what is Mabel doing then? Which hand does he turn it off with and does he take the towel before or after turning off the tap and with which hand?
FRED starts gingerly to dry his cut finger.	
MABEL opens the tin of dressings with her right hand and puts the lid down (upside down) on the draining board.	Note the action—particularly with the lid.
She takes out a dressing and puts the tin down on the draining board beside the lid (on the right side).	
FRED: Thanks love.	When does Fred speak? Is Mabel still taking the lid off the tin, or the dressing out of the tin or what?
FRED throws the towel down on the draining board.	Note that Fred speaks and then throws the towel down. Also note where it lands.
MABEL unpeels the dressing from its backing and places it on Fred's finger. Fred turns his head away and winces.	Does she use both hands to place the dressing on his finger? Note the way Fred's head turns to wince.
MABEL: Did I hurt you?	Note that she speaks before he turns his head back.
FRED turns his head back.	
FRED: No, it's just a bit sore.	
MABEL takes her hands away.	Note when she removes her hands.
FRED: Right. I'll finish the lawn.	

MABEL picks up the towel with her left hand and moves to the door.

Note the hand.

FRED picks up the lid of the dressings tin and places it on the box—picks up the box with left hand and moves to the door.

Note the hand, also that Mabel moved to door before Fred.

MABEL hangs the towel up at the door—FRED moves over to her with the tin which he hands over (his left hand to her right).

Note the hand action.

FRED: What time's lunch?

MABEL: Is half an hour all right?

FRED: Fine. Give me a shout.

Do not forget that during the dialogue Mabel is holding the tin.

FRED kisses her.
He pulls open the door and goes out.

Kisses her after the dialogue. Note. You see about the door?

N.B. when you come to do the scene directly following this one, which is Exterior, Garden, you will of course know that Fred must have a dressing on the relevant finger of his right hand, and that the back door is slightly open, at the start of the shot, before Fred comes out!

Remaining Coverage: Fred and Mabel Scene: 1

Having shot the master, the director goes on to shoot the rest of the coverage. This is your first test.

2-s Fred and Mabel at sink

From Fred turning on the tap—putting his hand under the water—you will remember which tap and which hand—Mabel comes up to him—she examines his hand—chat—tap still running—she replaces his hand under the tap—she exits frame (do you know which way?)—chat with her out of vision—she returns with towel and tin (in which hands?)—Fred turns off tap and takes towel—he dries his hand. Mabel meanwhile goes through the same actions as the master shot opening the tin. Fred puts the towel on the draining board. Mabel places dressing on Fred's finger. Fred turns his head away (which way?) and winces. Mabel speaks—Fred turns back to her and speaks—Mabel takes her hand away. Fred says he will finish the lawn. Mabel takes the towel and exits. Fred does his action with the tin of dressing and also exits.

MCU tap and hand

Hand in to tap—turns on water (which tap and which hand?). The finger, (suitably wounded) is placed under the water.

CU dressing going on finger

Mabel's hand holding dressing—places it on Fred's outstretched finger and presses firmly into place. (Make sure that Fred's finger is no longer wet, as he has just dried it with a towel.)

CU Fred wincing

Fred turns his head and winces. Mabel says her line out of vision and Fred turns back.

Other shots planned

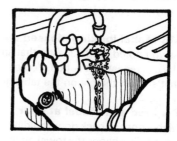

Tight two shot (Tight 2-s)

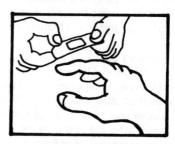

Medium close-up tap
and hand (MCU tap and hand)

Close-up plaster (CU plaster)

Close-up Fred (CU Fred)

Remaining Coverage: Fred and Mabel
Scene: 2

MCU cupboard

The cupboard door is shut to start. Mabel's right hand opens the door and reveals a tin of dressings and a bottle of antiseptic (both of which have been replaced after the W/A)—her right hand takes the bottle, then replaces it and takes tin of dressings—the cupboard door is left open and she moves her hand away.

MS Mabel looking pleased with the pie

As you noted on the master that she rubs her hands on the front of her apron, she can repeat that action. You also noted the way she turned to the oven, and that she left the pie on the table and did not take it with her.

2-s Fred and Mabel at door

Mabel goes to back door (holding the towel in left hand). The back door is half closed. She hangs the towel on the back. Fred enters from foreground to form a two-shot (Fred's finger is dressed and he carries the tin with him in his left hand). He hands tin to Mabel—chat—he kisses her and goes out through door.

Summing up

So, what are you left with at the end? First you have the whole scene played in wide angle, with everything contained in the one shot. Then you have a number of other shots from different angles of various sections of the scene. Because you have ensured that the action during every shot is as near as possible identical to the master wide angle, it will be possible to cut, say, from the master of the 2-s of Fred and Mabel at the sink, to the MCU of the taps, and so on at any one of a number of places. The cuts will then be determined in order that each piece of action may be seen in the final result from the most suitable angle in accordance with the dramatic interpretation of the script, and the resulting material available.

116

Remainder of shots planned

Medium close-up cupboard
(MCU cupboard)

Mid shot Mabel (MS Mabel)

Two shot at door (2-s at door)

When it's put together.

Fred and Mabel: Edited Sequence

Finally, the scene has been edited, and the completed sequence might look like this:

Scene 1: Int. Kitchen. Day

1. W/A   /MABEL IS PREPARING AN
2. MS Mabel   APPLE PIE/
3. W/A   FRED COMES IN/FROM THE GARDEN AND RUSHES
4. MCU tap   OVER TO THE SINK. HE TURNS/ON THE TAP AND HOLDS HIS HAND UNDER THE STREAM OF WATER
5. 2-s at sink   MABEL GOES OVER TO HIM/

   MABEL: Fred, Fred, what's happened?
   MABEL EXAMINES HIS HAND
   MABEL: How on earth did you do that?
   FRED: The bloody adjuster stuck on the lawn mower.
   MABEL: You've got dirt in it.
   SHE PUTS HIS HAND BACK UNDER THE TAP
   MABEL: I'll get you a dressing

6. W/A   SHE/GOES TO THE CUPBOARD AND GETS
7. MCU cupboard   OUT/A TIN OF DRESSINGS

   MABEL: Do you want some antiseptic on it?

8. W/A   /FRED: No, don't bother.
9. 2-s   MABEL TAKES A TOWEL AND GIVES IT TO FRED./ HE DRIES HIS HAND ON IT.

   FRED: Thanks love.

10. CU dressing   MABEL STICKS DRESSING/ON FRED'S
11. Cu Fred   FINGER/HE WINCES
12. 2-s   /MABEL: Did I hurt you?

    FRED: No. It's just a bit sore. (PAUSE)
    Right. I'll finish the lawn.

13. 2-s by door   BOTH GO/TO BACK DOOR

    FRED: What time's lunch?
    MABEL: Is half an hour all right?
    FRED: Fine. Give me a shout.

14. W/A   HE KISSES/HER AND GOES OUT.

The edited sequence

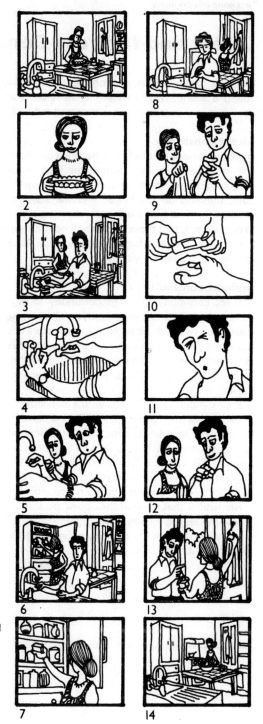

1. Wide angle
2. Mid shot Mabel
3. Wide angle
4. Medium close-up tap and hand
5. Tight two shot
6. Wide angle
7. Medium close-up cupboard
8. Wide angle
9. Tight two shot
10. Close-up plaster
11. Close-up Fred
12. Tight two shot
13. Two shot at door
14. Wide angle

119

Direct and Indirect Continuity

We have examined the relationship of action, dialogue, props and costume within a scene. Now that scene must be related to others. Two forms of continuity are involved here: direct and indirect continuity.

Direct continuity

Direct continuity means that actions are carried over from one scene to the next with no time lapse. Reverting to the Fred and Mabel scene, there was direct continuity between the end of Scene 1, with Fred walking out of the back door, dressing on his finger, and the start of Scene 2, with Fred coming through the door into the garden, dressing still on his finger. What is important to remember is that these scenes will not be shot consecutively—indeed Scene 2 may well be shot first, therefore accurate notes on the beginnings and endings of scenes are vital.

Indirect continuity

Indirect continuity refers to continuity links between scenes that are not consecutive. It means that there is some sort of break—a time lapse, a scene interposed which deals with other characters or some other aspect of the story, but where there are continuity links between these scenes however spaced out in the story order. These different continuity threads might not be immediately apparent on reading the script or when each individual scene is shot, maybe days or weeks apart.

For example: Scene 1 ends with a man stuffing a package into the right hand pocket of his coat. Scenes 2 and 3 are centred round another character. Scene 4 shows our man from Scene 1 walking along a street to a rendezvous. It must be remembered that he is wearing a coat from Scene 1. Scene 5 shows the man meeting his contact. The man hands over the package. There is therefore indirect continuity with Scene 1, the package in the right hand coat pocket; with Scene 4, wearing the coat; and possibly for the contact with later scenes.

Remember that none of the above are shot in story order, which makes the whole matter more complex.

Knowing the script thoroughly prevents you from being caught unawares by indirect continuity.

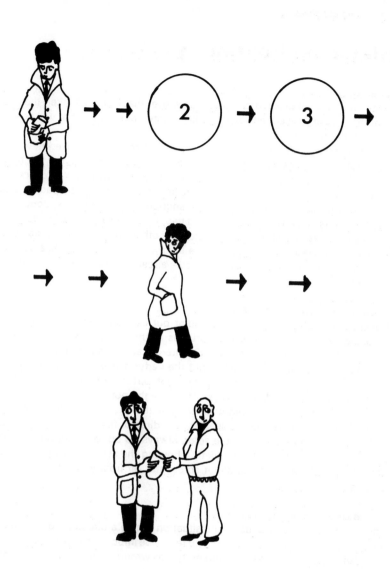

Scenes related to others

Scene 1 ends with the man putting the parcel into his coat pocket

Scenes 2 and 3 are concerned with other things

Scene 4 shows our man from Scene 1 (therefore there is indirect continuity from 1 to 4)

Scene 5 shows the man handing parcel over (therefore there is indirect continuity with Scene 1 and direct continuity from Scene 4)

121

Matching Continuity

Checking
Once you have established continuity on a shot that has been taken then any shot that, in the final edited version, will cut into or onto that shot will need to match for continuity.

In matching continuity, therefore, you are not so much involved in writing fresh notes but in checking. You check the artists' costume against your notes and photographs; you check the props against your drawings. You check that the camera angle is not going to cross the line so that objects and people jump in frame. During the rehearsals and the shooting you check the dialogue and that the actors' actions are the same and happen at the same time in relation to the dialogue as the shot already taken. In other words, you are not busy writing down fresh notes, unless there is new material being rehearsed and shot, but you are using those you have already made in order to match continuity.

If an actor gets it wrong
Never go directly to the artist and point out continuity errors without first speaking to the director. It is discourteous to the director, who is ultimately responsible for accepting the actor's interpretation and performance. The continuity might have been wrong but the performance so stunning that the director might not wish to retake the shot but decides to get round the continuity problem in other ways. There might not be time to retake for continuity or the director might know that the specific continuity error is unimportant. You must always speak to the director first.

If you are then asked to tell the actor, always go up to him or her with the script and explain the problem quietly, showing the actor the script. It is very confusing to the actor if you shout across the set from your position by the camera.

If the director won't listen
There is nothing you can do in the situation where a director won't listen or ignores a point of continuity which you feel to be important, except note on the continuity report sheet that the continuity was wrong. That could prove helpful to the editor.

Which 'take' do you match for continuity?
Let us say you have shot six takes on one particular angle. The director thinks that take 1 is possible and so is take 6. The continuity on those takes differs, and you are about to shoot another shot which has direct continuity with the one just recorded. You ask the director which take he prefers, in order to match continuity. The director will not, or cannot, specify, saying that he will look at both takes in the editing. Do you match continuity to take 1 or take 6?

As the director has not spoken in favour of either take, it is then up to you to decide which to match for continuity. I would always match continuity to the last take (take 6 in this example), as the artists are far more likely to remember how they performed five minutes ago rather than two hours ago. Also, the last take must have been satisfactory enough for the director to move on to another shot.

'If you are then asked to tell the actor, always go up to him and explain the problem quietly.'

Shooting on Videotape

For some reason, people often get confused about the necessity for continuity when recording onto videotape. The past legacy of videotape being thought of solely in terms of a multi-camera studio-based facility remains.

But videotape has, over the past ten years, changed enormously. The advent of videotape in cassette form has made a far more mobile and flexible resource than the studio or outside broadcast operation to which it was previously confined. Cameras have become smaller and lighter and receptive to far lower levels of lighting than film. The greatest changes have taken place in the area of videotape editing which, from very primitive beginnings, has progressed to the stage where, for example, most TV commercials, shot on film, are immediately transferred to videotape and edited electronically.

All this means that there is no hindrance to the flexibility of shooting on videotape using the film-style technique of single-camera, out of sequence shooting. However, once a film-style technique is adopted, the need for continuity is there, and the PA should be quite clear about her role.

Scripted

If the material is scripted then the role of continuity is just as important as if shooting on film. Sometimes a slate board will be used, not so much as a means of identification but rather as a discipline to alert the unit that recording is about to commence.

Timecode

The shot is identified by means of the 'in' point of timecode, and this must be noted either on a continuity sheet or a shot list. Timecode is a method whereby each video is identified by means of numbers broken down into hours, minutes, seconds and frames. Each frame always keeps its original identity, known as the timecode address. The timecode recorded onto the tape can either show the tape running time or the actual time of day.

Unscripted

If the material is unscripted then the role of continuity remains the same as for the same type of programmes shot on film. The shot list becomes additionally important, as tape is cheap in comparison to film and directors and cameramen will not stint themselves in the amount they shoot.

14 : 20 :55

Timecode logging
Timecode shown in hours, minutes and seconds.

Working on Documentary-type Productions: 1

You will always know something

Even if you are working on a production that appears to be totally unscripted and unplanned, you will always know *something* before you arrive on location—even if it is only the subject matter.

If, for example, you know nothing more than that you are working on a documentary about drink, then at least you can begin to focus your attention somewhere. Anything liquid, in bottles, restaurants, homes, on tables, in shops, should attract your interest.

Continuity

The job of continuity will, of necessity, be less precise than on scripted productions, but all the points in earlier chapters should be taken into consideration:

Shot description

An accurate and full shot description is of first importance. It is not enough to note simply 'Close-up large machine' without finding out and stating what the object is. This is especially important when the subject matter is of a technical nature. Neither is it sufficient to note 'Close-up of manometer' without giving a description. The editor is extremely unlikely to know what a manometer is and the director may well have forgotten.

Names and addresses

Always note down the names, addresses and titles of those taking part, e.g. *Chairman of Blogswich Cricket Club, 1951*, taking care to ensure that the spelling is accurate. Your note on your shot list may end up on a graphic on the completed programme.

Locations

Always note the location in your shot description. To state:

'W/A mountain

W/A different mountain

GV group of mountains'

is not that helpful to the editor, the director or the person writing the commentary.

'If you're working on a documentary about drink ... the job of continuity will, of necessity, be less precise.'

Working on Documentary-type Productions: 2

Costume

Costume details should be noted, especially if members of the public, dressed in their own clothes, are involved. If clothes continuity is important, then Mr Bloggs, whose geraniums won first prize at the flower show, must be tactfully asked to wear the same scruffy gardening clothes on three successive days of shooting. Where a presenter is seen in different locations, all shot out of sequence, then continuity of clothes should be carefully worked out and discussed in advance.

Basic continuity

Simple, basic continuity should be observed, such as exits and entrances and moves within a shot. Attention should be paid, as far as possible, to the conventions of single-camera shooting such as the avoidance of crossing the line and screen direction.

If a demonstration is given, with ad lib commentary, then the moves should be noted for subsequent cutaways.

Interviews

When an interview is shot, the questions should be noted down in order to relay them to the interviewee for later shots. It is helpful to note the timecode of each question if possible.

Public relations

One of your main tasks, when shooting, lies in the area of public relations. You may find yourself placating irate gentlemen who are unable to park their cars in their usual place due to the crew vehicles. You might spend hours explaining to members of the public what is going on and why there are lights and cameras. You will need a pocketful of coins to bribe children to go away, and you could find yourself chatting to nervous old ladies who have not had a wink of sleep the previous night, partly in dread of their forthcoming interview and partly for fear of disarranging their expensive hairdo, in a manner quite unlike your usual, rather reticent self.

In short, you act as trouble-shooter, paver of ways and smoother-out of difficulties.

The best advice I can give is to be prepared for anything to happen.

Trouble-shooter
Much of your job during shooting will be that of general factotum, fixer and trouble-shooter. You will smooth the way for an untroubled, fast, efficient shoot, and if anything goes wrong you will be blamed.

Writing It Down

The person doing the job of continuity needs to make a lot of notes. These are of use:

1. To yourself and the unit while shooting is taking place, in order to preserve continuity from shot to shot and scene to scene; and
2. To the editor, by providing a written log of the shooting.

Scripted and unscripted

If you are working with a script many of your continuity notes will be written on the script itself, as you will be able to pinpoint actions precisely in relation to the dialogue. You will, in addition, fill in continuity report sheets in order to provide shot-by-shot information. A fresh sheet should be used for every shot. There are various printed sheets available, which vary considerably in design and layout. These can be used when recording either on film or on videotape.

If you are working without a script you will need to provide a detailed shot list. I would use a notebook or larger sheets of ruled paper in order to compile the shot list on location.

How to organise the paperwork

When you are standing in a muddy field with the wind wreaking havoc with your notes and the director asks you whether, in Scene 64, the actor has his coat undone or done up, do you—after inwardly cursing—have to get down on your knees and wrestle with several large files in order to find the answer? How, in other words, do you organise the paperwork you carry with you on location in order to make the job as simple as possible?

Most people develop their own systems, after a certain amount of trial and error, and if you have one which works well for you, then don't consider changing it.

The system I use

I keep the script in a file in story order. I do not have another version of the script in shooting order as my interest is in the continuity between scenes when the material is finally edited together.

Each day I will extract from this file the pages of the scenes we are about to shoot. I put these on a clipboard, on top of blank continuity sheets which are held down by a rubber band. It becomes very easy, then, to work on the script and make notes on the continuity report sheet by flicking from one to the other.

At the end of the day I will replace the pages of script in the file in story order together with my rough continuity sheets and any relevant photographs. In this way, any query about Scene 64 only means looking in one place in one file.

PRODUCTION			EPISODE	SCENE	SLATE / IDENT
NUMBER		DATE			

FILM CAMERA ROLL	FILM SOUND ROLL	VIDEOTAPE REEL NUMBER	(Circle whichever is appropriate)					
			INT.	DAY	NIGHT	SYNC	GUIDETRACK	WILDTRACK
			EXT (Weather)			MUTE	PLAYBACK	

LOCATION / SHOT DESCRIPTION

CAMERA INFORMATION (If appropriate)
LENS _____
DISTANCE _____
STOP _____
FILTERS _____

TIMECODE OR ENDBOARD	TAKE*	DURATION	REMARKS (Reason if take is NG)	CONTINUITY NOTES (Including costume/props/dialogue/drawings etc.)
	1			
	2			
	3			
	4			
	5			
	6			
	7			
	8			
	9			
	10			

* Circle acceptable takes - if more than 10 takes, use second sheet.

© Avril Rowlands 1989

When You Make Notes

You are standing on the set, by the camera, on the first day of shooting. The world seems to have gone mad around you, with everyone busy rushing round, attending to their own concerns. At what point do you start to do your job and begin taking notes?

Write what you can when you can

You should start to take notes as soon as possible. The order in which things will happen will vary, but you should make it a rule to write down what you can when you can and always be alert to what is going on around you.

If you are working on a production where continuity is of great importance, you should *never* leave the set when shots are being discussed, set up, rehearsed and recorded. You should not retire to a little room, set aside for your purposes, in order to type up yesterday's continuity sheets. If you do, you will not be able to keep up with what is happening today.

Although it would be helpful to watch a rehearsal without taking any notes, you might find that you then don't have enough time to take sufficient notes later on. You will never be able to write down all you want and you should use every second available.

Pens or pencils?

Do you make notes using pencil or coloured pen? There are two options open to you, both with advantages and disadvantages.

First, always take rehearsal notes in pencil. That way, you can alter your notes as things change. If you continue to use pencil when shooting, you should make sure that you can distinguish the different takes, perhaps by writing in 'take 1' and 'take 2' beside the notes. The advantage of continuing to use pencil is that you can erase wrong or altered notes and your script remains fairly clean and legible. The disadvantage lies in being unable to differentiate between rehearsal notes and notes relating to the recording, unless you are very careful.

You will not have this problem if you use coloured pens for your notes when recording. The difficulty with this system lies with the director who shoots a number of takes of each angle. You will need a good colour-code system to remember that pink relates to take 12, and the danger is that your script will become excessively messy and hard to read.

132

'You should not retire to a little room, set aside for your purpose.'

Continuity Report Sheet: General

Many companies have their own continuity report sheets. The form shown opposite was designed by the author to be used on either film or video shooting.* Let us examine in detail what should be written on a continuity report sheet.

Production details
The first details to be written are the name of the production, the production number and date.

Episode/scene
The episode (if a series) and the scene or sequence number should be marked. A script is divided into scenes—each new location being a new scene. A scene is often called a sequence but, to be strictly accurate, a sequence is the final cut version of the scene. If there is a separate shot number, put that down.

Slate/ident
The slate number relates to material shot on film and is the number given to each photographed shot. It is a means of cataloguing the shot. Almost always the first shot of the first day's shooting is called Slate Number 1 and proceeds consecutively until the last shot of the last day of shooting. The slate number is shown marked up on a clapperboard at the start of each shot. This board has a twofold purpose. In the days of silent pictures each shot would be identified by the humblest member of the camera department holding up a piece of slate with a number chalked on it. When talkies arrived he would, in addition, bang together two pieces of wood—called a clapstick—to establish synchronisation between action and sound. Later the two were combined to form the clapperboard. The board is still called the 'slate' by tradition and the individual shot itself, being identified by a slate number, is colloquially referred to as 'a slate'.

If shooting on videotape a clapperboard might still be used to identify the shot, although there would be no need to bang together the pieces of wood. In this box, therefore, the shot number should be written. If there is no shot number then the 'in' point of timecode would act as the reference point on the tape for that particular shot. With the increasing introduction of timecode in film shooting for both synchronisation and identification, it is now possible to shoot film without using a clapperboard, logging the shots by timecode numbers in the same way as on videotape shooting. There are an ever-increasing variety of systems—Keycode, Aatoncode, Arricode—and, depending upon the system used, the timecode may or may not be the same for the pictures shot on a film camera and the sound, recorded with timecode either on a quarter-inch machine or a DAT recorder. If a clapperboard is used it may or may not incorporate a timecode display of its own.

I cannot over-emphasise the need for discussions between the PA, cameraman, recordist and editor before the shoot, to establish the most efficient way of logging the material.

Interior or exterior
Mark whether you are shooting an interior or an exterior scene and note the weather, if exterior. Just 'dull' or 'sunny' will suffice.

*Supplies available from the author, on (01386) 792 051.

FILM AND VIDEO CONTINUITY NOTES

PRODUCTION					EPISODE	SCENE	SLATE / IDENT
NUMBER		DATE					

FILM CAMERA ROLL	FILM SOUND ROLL	VIDEOTAPE REEL NUMBER	(Circle whichever is appropriate)						
			INT.	DAY	NIGHT	SYNC	GUIDETRACK		WILDTRACK
			EXT (Weather)			MUTE	PLAYBACK		

LOCATION / SHOT DESCRIPTION

CAMERA INFORMATION (If appropriate)
LENS _____
DISTANCE _____
STOP _____
FILTERS _____

TIMECODE OR ENDBOARD	TAKE*	DURATION	REMARKS (Reason if take is NG)	CONTINUITY NOTES (Including costume/props/dialogue/drawings etc.)
	1			
	2			
	3			
	4			
	5			
	6			
	7			
	8			
	9			
	10			

* Circle acceptable takes - if more than 10 takes, use second sheet.

© Avril Rowlands 1989

135

Continuity Report Sheet: Technical

Roll numbers
Note down the camera and sound roll numbers if shooting on film. If shooting on videotape, note the tape reel number.

Sound details

Stereo
With the almost universal use of stereo sound in finished films and television programmes, the PA should ideally log what sort of sound is taken for each shot.

M & S
In a stereo production, this could be M & S, which stands for Mono and Stereo—a system whereby one track on the recording machine is used to record a mono sound signal, and the other to record a stereo or width signal. The two signals are combined in post production to produce a final A + B stereo sound track.

A & B
Or the sound could be recorded A & B, which means basically left and right stereo as finally heard by the viewer on their television set.

Mono
Sometimes on a stereo production, the recordist will shoot the sound mono to achieve clarity, for example, in dialogue. This will be combined in post production with a stereo atmosphere, either recorded on location as a wildtrack or taken from a library effect.

Dual mono
In some situations the location recordist may decide to record twin track mono sound (not to be confused with stereo even in a stereo production). A good example of this is a conversation between two people each of whom is wearing a personal radio microphone. It is frequently better for the voices to be combined in post production by the dubbing mixer than mixed on location.

Shooting on film
If shooting on film, the next thing to note down relates to sound. Is the shot taken sync—that is, with camera and sound 'locked' together and running synchronously? Or is it to be a silent or 'mute' shot (with no sound recording being made). If the sound is being recorded without shooting a picture, this is known as a 'wild track', sometimes written as 'wildtrack'. The expression 'wildtrack with camera' means that the sound is recorded simultaneously with the shooting of the picture but with no interlock

between camera and recorder. All under- and over-cranked shots which have sound recorded for them at the same time fall into this category.

'Guidetrack' indicates that the camera and sound are running in sync: the sound is intended only as a guide to the editor and is not of good enough quality to be used in the completed film.

'Playback' is the term used where sound is played into the shot as a guide for the actors' movements. In a pop video, for example, an actor might mime the song to playback.

Shooting on videotape
If shooting on videotape, the camera will not be run at varying speeds. However, although a sound track is automatically recorded on videotape, it is still possible to have a mute shot—in other words there is no sound on the soundtrack. Conversely, it is still possible to record a wildtrack. However, as it is necessary to record some sort of picture for technical reasons when making 'sound only' recordings on videotape, it is immensely helpful to the editor if a shot, however rough, is taken of the microphone while the wildtrack is being recorded, as it visually identifies the sound as being a wildtrack.

Continuity Report Sheet: Camera Information

This information is sometimes given to you to write down when shooting on film, and on single camera video—especially when it is a drama shoot.

The information, when given, will serve an important function for the technical continuity of shot size, aperture, colour temperature and filters used during a shooting sequence—especially with drama when retakes or additional shots might be necessary and taken at a later time.

The information you may be required to write down refers specifically to the lens—both with film and video. A film camera will be equipped with a series of fixed focal length lenses—referred to as prime lenses—and a zoom lens, which usually includes most of the angles covered by the prime lenses. Additional lenses, such as a telephoto or narrow angle/long lens or a wide angle lens might be needed for specific shots.

A video camera will be equipped with two zoom lenses—a standard zoom lens and a wide angle zoom lens. Prime lenses for single camera shooting are also available.

For continuity of a shot size, you will be asked to write down the lens angle and lens distance of the shot. The lens angle is the angle of view of the subject as 'seen' by the lens and this information is of no use without the 'lens-to-subject' distance being written down as well in order to match shot sizes.

For continuity of exposure, you will be asked to write down the lens stop. The stop relates to the amount of light allowed through the lens to expose the film or videotape. On particular occasions the cameraman will deliberately 'under expose' the shot, i.e. make it darker—and will need to be reminded at a later date of the extent of under exposure to enable exposure continuity. With film, the stop—or lens aperture—is measured using a light meter, and with video, this is measured by the lens itself.

For continuity of colour of shot, you will need to write down the colour correction filter used for film, or the colour temperature for video. When using film, which can be balanced to artificial light or daylight, a colour-correction filter is used to convert the colour of the light being used to the colour for which the film has been pre-set. Tungsten film is balanced for artificial light. Daylight film is balanced for daylight. Sometimes tungsten film is used in daylight and daylight film in artificial light. In order to do this, the cameraman uses colour correction filters—an orange '85' when shooting daylit scenes on tungsten film.

These filters are not necessary when shooting on video, as the camera is able to balance for any colour of light. But to match colours in a shooting sequence he may ask you to write down the colour temperature of the light used. Artificial light would normally be about 3500 K, and daylight about 5600 K. The symbol K denotes the colour temperature in degrees Kelvin.

For continuity of effect, the cameraman has at his disposal a vast array of filters to create mood, style or emphasis of lighting. A film cameraman

138

will normally use such filters for effect only. A video cameraman uses some filters which are permanently attached to his lens to create a 'feel'. The filters are permanently in place but may change in a shooting sequence according to the subtlety of the 'feel'. A range of filters commonly used for video shooting to create a feel are the low contrast or LC. These are graded 1 to 6 and because the filter is always used, a subtle change of grade may be required. This will be essential information for the cameraman to retain in order to maintain continuity of shot 'feel'.

FILM AND VIDEO CONTINUITY NOTES

PRODUCTION LOST DAYS IN ATLANTA			EPISODE 4	SCENE 34	SLATE / IDENT 68
NUMBER 2T/1039/Y	DATE 30/4/94				

FILM CAMERA ROLL	FILM SOUND ROLL	VIDEOTAPE REEL NUMBER	(Circle whichever is appropriate)		
12	20	\	INT. (DAY) NIGHT	M . S (SYNC) GUIDETRACK	WILDTRACK
			(EXT) Overcast, rain (weather)	MUTE PLAYBACK	

LOCATION / SHOT DESCRIPTION

CAMERA INFORMATION (If appropriate)
LENS 25 mm
DISTANCE 12 ft
STOP 3.5
FILTERS 25
Col. TEMP K 5,600

TIMECODE OR ENDBOARD	TAKE*	DURATION	REMARKS (Reason if take is NG)	CONTINUITY NOTES (Including costume/props/dialogue/drawings etc.)
	1			
	2			
	3			
	4			
	5			
	6			
	7			
	8			
	9			
	10			

* Circle acceptable takes - if more than 10 takes, use second sheet.

© Avril Rowlands 1989

139

Continuity Report Sheet: Continuity

Next you write down the notes that are specifically for continuity.

Location
You should note the location, either by writing 'The Happy Haymaker's Pub, Grove Road, Dorset', or just 'pub'. It all depends on whether there is a separate list of locations on the schedule and how important you feel the information to be.

Shot description
This description should state how a shot begins, how it ends and any development in the middle. It should be as concise as possible and an easy reference for yourself or the editor.

Continuity notes
The space on the right of the sheet is left blank for you to fill in information which is important for continuity.

At the start of a new scene always write what the artistes are wearing and how their hair is arranged. Make a note of any props they are carrying, e.g. a walking stick. Notes about make-up are only made for a major effect, like a cut cheek, which would heal during the progress of the story.

You might use the space to note any specific dressing props, 'N.B. Fire must be alight', and you should make a sketch of the shot, with the position of the camera and actors noted down. You might also draw diagrams of set dressing—for example, a table laid for a meal.

Either note down the 'in' and 'out' words of dialogue or mark the coverage on your script. It is very important to know where in the scene the shot begins and ends.

FILM AND VIDEO CONTINUITY NOTES

PRODUCTION		EPISODE	SCENE	SLATE / IDENT
LOST DAYS IN ATLANTA		4	34	68
NUMBER ZT/1039/Y	DATE 30.4.94			

FILM CAMERA ROLL	FILM SOUND ROLL	VIDEOTAPE REEL NUMBER	(Circle whichever is appropriate)					
12	20		INT.	(DAY)	NIGHT	H.S		
						(SYNC)	GUIDETRACK	WILDTRACK
			(EXT) Overcast, rain (Weather)			MUTE	PLAYBACK	

LOCATION / SHOT DESCRIPTION

HIGH STREET

CAMERA INFORMATION (If appropriate)
LENS ___
DISTANCE ___
STOP ___
FILTERS ___

2-s Mr . Mrs Jones → they exit R at end

TIMECODE OR ENDBOARD	TAKE*	DURATION	REMARKS (Reason if take is NG)	CONTINUITY NOTES (Including costume/props/dialogue/drawings etc.)
	1			
	2			
	3			
	4			
	5			
	6			Mr J : black suit, jacket undone, red waistcoat, white shirt, red/white spotted cravat, hat r. hand / w. stick L. hand.
	7			
	8			
	9			Mrs J : green/grey striped dress, shawl (grey) over shoulder. Hair tied back in bun. Basket r. hand.
	10			

* Circle acceptable takes - if more than 10 takes, use second sheet.

© Avril Rowlands 1989

141

Continuity Report Sheet: Details of Shooting

The information on the left-hand side of the sheet relates to the shooting and is of vital importance to the editor.

E/B or T/Code
An endboard usually relates to shooting on film and refers to the clapperboard. Sometimes the board is not put on the front of the shot but at the end—for example, when something should be filmed instantly and without the delay of putting a board in front of the camera. The board is held upside down to indicate to the editor that it refers to the shot that has just ended.

A timecode reference should be used against each take when shooting either on videotape or on film when timecode is used on location without a clapperboard. Either the 'in' timecode of this take or the 'out' point of the preceding one should be shown.

Takes
A take is the recording of the action of a shot. If that action is not satisfactory for some reason the same action is repeated using the same slate number (if shooting on film) or shot number, but identified as 'take 2'. If that is not satisfactory either, there is another take, and so on until the director is satisfied.

On your continuity sheets you should circle the accepted takes. The others are marked NG (No good). But you should always give a reason *why* a take is NG. This is most important for the editor. Give as full a reason as possible. The take might have been good until the end, when the noise of a car intruded. Write that down. The director might have been dissatisfied with an actor's performance or the cameraman with his framing. Write down the cause, whatever it is.

False starts
The director calls 'action' to start the shot but after a few seconds the actor might fluff his lines. Instead of 'cutting' the camera and starting again with a new take number the director might tell the actor to start again while the camera is still running. Make a note of that false start.

Duration
Give a timing for each take, from the moment the director says 'action' to when he says 'cut'.

PRODUCTION		EPISODE	SCENE	SLATE / IDENT
Roses are Red		\	10	42
NUMBER 34907	DATE 1.5.94			

FILM CAMERA ROLL	FILM SOUND ROLL	VIDEOTAPE REEL NUMBER	(Circle whichever is appropriate)				
\	\	7	INT. DAY NIGHT	SYNC GUIDETRACK	WILDTRACK		
			EXT (Weather)	MUTE PLAYBACK			

H - S

LOCATION / SHOT DESCRIPTION

W/A living room (Roger's house)
→ tracking R

CAMERA INFORMATION (If appropriate)
LENS _____
DISTANCE _____
STOP _____
FILTERS _____

TIMECODE OR ENDBOARD	TAKE*	DURATION	REMARKS (Reason if take is NG)	CONTINUITY NOTES (Including costume/props/dialogue/drawings etc.)
11.02.23	1	33"	Performance	(A) - (K)
11.03.15	2	30"	Light in shot	Costume notes - see 40
11.40.10	3	35"	OK A-C D-K NG-dial.	N.B fire lit, ditto candles
11.05.05	(4)	32"		
	5			
	6			
	7			
	8			
	9			
	10			

* Circle acceptable takes - if more than 10 takes, use second sheet.

© Avril Rowlands 1989

Writing on the Script

As well as making notes on the continuity report form you should use your script in noting down continuity points. The great advantage of writing on the script lies in the fact that you can match the action to the dialogue with great accuracy.

Abbreviations of words and the use of symbols will enable you to write quickly on the script. Some people use a form of shorthand.

Below are a few commonly used abbreviations and symbols:

CR   Camera right
CL   Camera left
R   Right—from the actor's point of view
L   Left—from the actor's point of view
     (You might prefer to use the abbreviations R and L for camera right and left and r and l referring to right and left from the actor's point of view.)
X   Could be used to denote 'crosses', i.e. Fred X R—Fred crosses camera right, or to denote 'exits' or 'extra', i.e. XXR—extra exits camera right.
f/w   Forwards
b/w   Backwards
i/f   In front
b/h   Behind
f/g   Foreground
b/g   Background

An arrow can be used to denote a turn, a stand or a sit. ⤵ ↓

A stick man can be used to show arm movements. This is very useful when you want to note whether someone is eating, drinking or smoking.

Pause   Hands together   Arms crossed   Noses

Dialogue changes

It is part of your job to keep a note of dialogue changes, and you should keep your script up to date as these changes occur.

SCENE 505 : EXT. STREET
PAULINE WALKING

SCENE 506 : DAY. INT. OFFICE

LAURA IS SITTING BEHIND HER DESK, WHICH
IS LITTERED WITH PAPER AND
PHOTOGRAPHS. EVEN THE CHAIR IS FULL
OF PAPERS. THERE IS A KNOCK ON THE
DOOR.

Writing r. hand
looks up

LAURA

X Come in!   *looks down. Carries on writing*

P top held
On   arm

P × R

PAULINE ENTERS THE ROOM AND
APPROACHES THE DESK

PAULINE

I've come about the advertisement. *top to shoulder*

LAURA *(head down)*

Just a minute. (SHE FINISHES WHAT SHE IS
DOING BEFORE LOOKING UP) You've come
about...?

PAULINE

advertised

The advertisement. You know, the one in the
paper.

LAURA *↑ stands*

P. top to floor ———— Oh, of course! (SHE SHAKES HANDS) Do sit
down. Just put that junk anywhere. ~~I always work~~   *L side. Picks up pen*
~~in an awful mess, don't you?~~ Now you're...?

Other Paperwork

Daily continuity log
A daily log should be kept of the shooting, in shooting order. This should consist of the roll numbers, the slate number or timecode, the scene and the takes. This could prove very useful if you need to refer to a specific shot and only know that it was Slate 63.

Daily progress report
In the feature-film industry the production office will require a daily progress report to be compiled by the person doing continuity. These are more or less complex documents (the example given opposite is a simple version), which can include a comprehensive breakdown of every minute of time spent on location.

The progress report provides the production office with information on the times of shooting, for possible overtime payments, together with details of time spent in meals, in setting up and in any technical holdups. It also shows how much edited material has been shot each day, comparing it with the overall running time of the film, the amount already shot and the amount left to do, as well as giving the number of scenes covered each day, any retakes and any wildtracks.

DAILY CONTINUITY LOG

Date: 3.5.94

Production No: 6581

Title: 'THE BROWNING AFFAIR'

Slate	Take	Scene	Cam. roll	Sound roll
324	1	45	16	9
"	2	"	"	"
"	3	"	"	"
325	1	46	16	9
326	1	12	16	9
"	2	"	17	"
"	3	"	"	"
"	4	"	"	"

DAILY PROGRESS REPORT

Date: 3.5.94

Production No: 6581

Title: 'THE BROWNING AFFAIR'

SHOOT CALL: 0800
1st shot: 0915
LUNCH: 1255
1st shot: 1445
WRAP: 1835

	PAGES	SCENE	MINUTES
Total script	64	40	30.00"
Shot previous	20	15	10.15"
Shot today	5	1	3.15"
Total to date	25	16	13.30"
To do	39	24	16.30"

Scenes covered	Retakes	Wildtracks
6	-	14X - 20X

Information For the Editor

There are two ways of providing the editor with all the information he requires when working on a scripted production.

Typed-up continuity report sheets

You can type up all your rough notes on to a continuity report sheet and give it to the editor. That is the system most people follow and one which I used myself for many years until I happened to mention to my husband, who is a film and videotape editor, that all the additional typing in hotel bedrooms late at night or early in the morning were, for me, the one big drawback to the job.

His reply surprised me, for he said that he did not need half the information which had been so painstakingly given to him. For example, he was not interested in knowing what costumes the artists wore—provided that continuity of costume was correct from shot to shot and scene to scene. Likewise the full description of the shot did not interest him. He did not need to be told that Jim entered shot frame L—walked R to the table—sat down—picked up the full glass of beer (with his right hand) etc. He had only to look at the relevant piece of film or tape to find all that out!

So together we devised a far simpler way of providing the editor with the information he really does need.

Continuity cards

The cards contain only the barest information about the shot, but it is information which the editor really needs. He needs to know which scene the shot relates to—the slate number or 'in' point of timecode, the camera or tape roll number and the sound roll number. Make a note if the take is sync or mute and use a separate card for wildtracks. Apart from that, he wants a brief description of the shot, e.g. W/A Henry's kitchen, and then, and vitally important to the editor, he needs to know about the takes: how many there were, which were NG and why, and which were acceptable. He needs nothing else *except* an accurate marked-up coverage script to tell him what the shot covers in the scene.

The cards are simplicity itself. You will find that you have time during the day to write them out, reasonably clearly and legibly, and if you are working on a series or inserts for a series, I find that different coloured cards corresponding to the colours of the scripts are useful.

Having gone on at some length about the virtues of using cards, it is only fair to say that some editors do not like working with cards—they prefer pieces of paper which they can then place in a file. There is, of course, no reason why the basic information the editor needs should not be written on pieces of paper rather than cards. The important point is to give the editor the information he wants and not a lot of extraneous detail which he will never look at, but which has caused you hours of unnecessary work.

Continuity sheet for the editor

Title ....

....

Date ....

EPISODE	SCENE	SLATE

Camera roll*	Sound roll*	Circle whichever is appropriate:
		SYNC GUIDE PLAY MUTE WILDTRACK
		TRACK BACK (to cover)

SHOT DESCRIPTION

Take*	Duration/ footage	Remarks (give full reason if take is NG)
1	* circle accepted takes	
2		

Continuity card

Slate	Ep/Scene	Neg. roll	Tape roll

Description

Take	1	2	3	4	5
Duration					
Remarks					

Coverage/Tramline Scripts

If you use the card system of continuity then it really is very important that you give the editor an accurate marked-up coverage script together with the cards so that he can see what the shot covers in the scene.

Even if you do not use the card system, but type up continuity report sheets for the editor, he will find coverage scripts useful and you and the director might also find it helpful to have a marked-up script for yourselves.

On a copy of the script you should draw lines for each shot showing the total coverage of the scene. Identify each line by using different colours and write the slate number or 'in' point of timecode alongside, together with a brief description.

Do *not* draw a line for each take, just for each shot. Draw the lines on the left-hand side of the script, keeping them clear of the dialogue and mark any dialogue changes in the copy for the editor.

If this coverage script is accurate it will be of immense value to the editor and will also show you and the director the coverage of any scene at a glance.

Don't confuse this coverage (sometimes called a 'tramline') script with a possible cutting order which the director might provide for the editor.

Example of coverage script

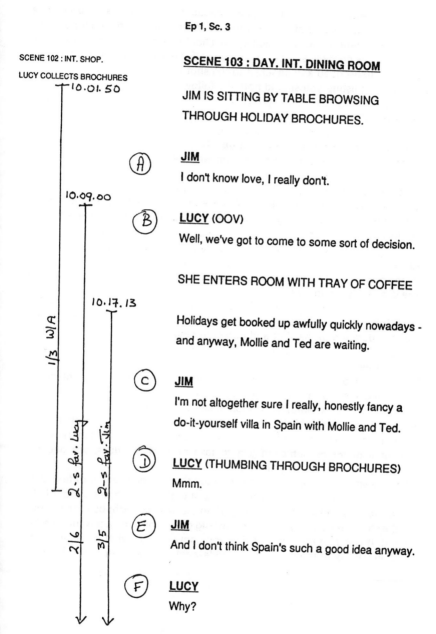

SCENE 102 : INT. SHOP.

LUCY COLLECTS BROCHURES

10.01.50

10.09.00

10.17.13

1/3 W/A

2-s fav. Lucy   2-s fav. Jim

2/6   3/5

SCENE 103 : DAY. INT. DINING ROOM

JIM IS SITTING BY TABLE BROWSING
THROUGH HOLIDAY BROCHURES.

(A)

JIM
I don't know love, I really don't.

(B)

LUCY (OOV)
Well, we've got to come to some sort of decision.

SHE ENTERS ROOM WITH TRAY OF COFFEE

Holidays get booked up awfully quickly nowadays -
and anyway, Mollie and Ted are waiting.

(C)

JIM
I'm not altogether sure I really, honestly fancy a
do-it-yourself villa in Spain with Mollie and Ted.

(D)

LUCY (THUMBING THROUGH BROCHURES)
Mmm.

(E)

JIM
And I don't think Spain's such a good idea anyway.

(F)

LUCY
Why?

151

Shot Listing

If you are working on an unscripted production the shot list you compile during shooting will be of vital importance to the director, the editor and the person who writes the commentary. It should therefore contain as much information as possible and be typed out neatly after shooting. Any interviews should be transcribed and attached to the shot list.

When shot listing you should note the following:

Location

Slate number or 'in' point of timecode
Give these details together with the number of takes.

Shot description
Make this as concise as possible while describing the shot accurately. Always write down the different stages of a developing shot and note especially how the shot begins and ends.

Sound
When shooting on film always note whether the shot is taken sync or mute and note down any wildtracks.

Duration
Time the duration of the shot and note that down.

Interviews
At some stage or other you will almost certainly be shot listing for an interview. There are one or two things to note about it.

Unless you are extraordinarily good at shorthand, or the interviewee speaks very slowly, you are unlikely to get it all down on paper. You can either take a tape or cassette recorder and make a rough recording which you can transcribe later, or it might be sufficient just to take note of the questions and the beginnings and ends of the answers.

The questions are important, as very often the interviewee is shot first, then one or two 'noddies', as they are called, are shot. These are shots of the interviewee, not talking, but listening. Then the camera might be turned round and the shots taken of the interviewer asking questions, with a few 'noddies' at the end. It is important therefore that you know exactly what questions were asked. Finally, or at the beginning, a wide-angle establishing shot is taken.

SHOOTING ORDER SHOT LIST

'STATELY HOMES'   Director: Peter McDonnell   21 May 1994

Programme No: 21Z/367942/B   Location: Withyton House, Staffs

T/C IN	TAKE	SHOT DESCRIPTION	DUR:	OK/NG
10.04.00	1	**EXT: HOUSE & GROUNDS** Wide establishing shot of house from gates - looking down tree-lined drive	45"	OK
10.12.05	1	z/o from house to WS a/b	1.00"	NG - person in shot
10.16.20	2	a/b	1.05"	OK
10.32.45	1	Start wide & z/i to house	20"	NG - cam
10.43.40	2	a/b	55"	OK
11.55.10	1	**EXT: SUNKEN GARDEN** WS panning L-R to gazebo	40"	NG - cam
11.57.28	2	a/b	42"	OK
		INT: DRAWING ROOM **INTERVIEW WITH LORD & LADY CORLEY**		
12.14.48	1	2-s Lord and Lady C on sofa	6.40"	OK

Q: Lord and Lady Corley, this beautiful house is an
amalgamation of many different styles of architecture. When
was the first building constructed?

A: (Lord C) There has been a house on this site since
Elizabethan times...bits added by each succeeding
generation.

(14.39.00)   Q: And have you added anything?

153

The Realities of it All

Location work can be very tough. You will undoubtedly feel many times that there are easier ways of earning a living as you stand half-frozen and soaked to the skin in a field at three in the morning trying to shoot a group of sheep who refuse to cross the field from left to right. Or when you are lost in a mist on the hills with a director who has lost the map, lost his and everybody else's way, and, even more important, lost the location caterers who are somewhere down in the valley.

Continuity itself is hard work. It requires an immense amount of concentration, not only when shooting is actually taking place, but when nothing is apparently happening and most of the unit appear to be idling around. Most of them *are* idling around, but if the director and cameramen are working, then you must be working too.

Continuity also involves sitting up typing and marking up editing scripts often into the small hours while everyone else relaxes, enjoyably fraternising with the local populace.

If, however, you feel I have drawn too black a picture, do not forget that if you *do* prove to be one of those strange people like myself who love the work, and find continuity a most satisfying job, you will find like-minded people on every new unit.

And the discomforts? . . . Well they do make for good after-dinner stories.

'. . . you see, there I was, standing quietly watching the action, when I felt something tugging my hand. And it was this goat we'd borrowed from the local farmer to use as dressing. You know what it was up to? It was quietly eating my continuity notes in a contented sort of way . . .'.

1

2

3

4

The realities of shooting
1. Shooting can be frustrating
2. It can be annoying—when the director has lost the way
3. It can be hard work—typing at night when everyone else relaxes
4. But it does provide good after-dinner stories

Post Production

It is very easy, at the 'end-of-production party', to think that all the work is over. The artistes are leaving, their minds on their next role, or lack of one; the crews are busy thinking and talking about their next job. The sets are dismantled and the costumes and props returned to the hire companies. The close-knit unit, so like a family in its internal working, is splitting up and there is a general air of relief at a job completed and the thought of returning to civilisation, tinged with sadness at the ephemeral nature of the business.

You prepare the paperwork for the editor and, apart from tidying up loose ends, might also feel that the production is over. But for many people the end of shooting signifies the start of their job. Unless editing has been taking place at the same time as shooting, the work of the editor starts when your job ends.

Editing
What has been recorded on film or videotape is a mass of unrelated material, shot out of sequence. It is now the job of the editor, working creatively with the director, to reassemble this material into its final story order. The picture will be edited, the sound mixed together and captions or graphics added before the production can be called complete.

Principles of editing
The mechanics of editing, i.e. the equipment used and the application of that equipment, will differ greatly according to whether film or videotape is being edited, but the basic principles of editing remain the same.

The editor's role is to juxtapose visual images and sound creatively and dramatically so that a story is told in sound and pictures.

'The editor working creatively with the director.'

Editing the Picture: 1

It is not necessary for PAs to understand the various possibilities for post production and some may find the following chapters hard to follow. I have included them because I am increasingly being asked questions about post production on my training courses and as many PAs now have to book the editing facilities I hope these chapters may be helpful. Unless one editor is going to be in charge of the whole process both off-line, on-line and track-laying, it is important to ensure that everybody involved talks to one another and agrees on the technical route to be taken before shooting starts. Otherwise, in an extreme case, it would be possible to arrive at a perfect off-line edit which cannot be related to the master material.

The choice of editing method used for a production depends upon a number of factors: location format, delivery or transmission format, access to particular sorts of equipment, and previous experience of editors and producers all have a bearing on the choice of post production method. In its simplest form, if a programme has been shot on tape, the editing might only involve an on-line edit where the editor re-recorded the pictures and sound shot on location in the right order on another role of videotape. At its most complex, a production shot on film might have its negative transferred on to videotape, before using a non-linear off-line editing system to create three separate versions of the material. Using the videotape masters, two versions might be on-line edited to form two television programmes of different lengths, while, by means of computerised negative cutting lists, the original master negative might be physically cut to form a version for cinema release.

So what are all these processes?

Conventional film editing
When you record on videotape the picture and sound are on the same tape or cassette and you can play them back straightaway. When using film, the pictures are shot by the camera onto film stock and the sound is recorded on a separate tape recorder onto quarter-inch wide magnetic tape or DAT. The sound can be played back immediately, but the picture cannot until the exposed film has been processed at a film laboratory.

Before any editing can take place, the sound has to be re-recorded onto magnetic sound film, of either 16mm or 35mm width, usually according to the gauge used for the picture. Once this is done, the picture and sound are synchronised by the film editor's assistant then viewed by the producer, the director and the film editor so that the final selection of takes may be made. After viewing, the editor breaks down the rolls of picture and sound, physically cutting them up into individual shots for easy access and the editing process begins.

The editing is done by joining these shots together in whatever order is decided upon by the director and editor. Shots can be experimented with in any order, and they can be lengthened, shortened, replaced or re-cut at any point. When using negative/positive film stock it does not matter what state the cutting copy gets into: it is only a guide for the negative cutter who splices the negative only after all the creative decisions have been made. This system allows immense flexibility. It means that changes can be made at any time until the negative is cut. Once the negative is cut, the show print or prints can be made without loss of picture quality, since they are made directly from the picture negative.

Editing the Picture: 2

Videotape editing

At first sight videotape editing seems to be simpler than film editing. Sound and picture are already in sync on the same physical piece of tape, and all the editor has to do is to re-record the sections wanted in the right order on to another role of tape. This gives us pictures and sound of second generation since to go back and make alterations means going to a new generation of tape. Each time material is copied on to a new generation of analogue videotape you lose both picture and sound quality. The quality lost in a digital edit suite is imperceptible, but the cost of on-line digital editing can be prohibitive. Thus most complex programmes make use of off-line editing.

Off-line editing

The essential point about off-line editing is that the master material is transferred to a lower quality, and often more flexible format, while the creative decisions are made about its order. This has the advantage of: (a) not wearing out the master material; (b) reducing the cost of 'thinking' time by not tying up expensive on-line equipment; (c) allowing the process to be more flexible, particularly with non-linear editing systems.

Off-line video editing originally supposed that the master material was in the form of time-coded rolls of videotape which would eventually be edited almost automatically in the on-line edit using an Edit Decision List (EDL) derived from the off-line. Now, with the advent of timecode in film shooting, it is possible to perform an off-line linear or non-linear edit and, by means of computer programs, produce a negative cutting list by which the negative cutter cuts the film negative in the conventional way without the film editor having used the conventional film editing process at all.

Linear off-line editing

This involves transferring the broadcast quality master videotapes on to a lower quality cheaper videotape format with matching timecodes. Likely formats are VHS, S-VHS, High or Low-Band U-Matic, and the 'Universal' range of low-cost Betacam SP VTR. Although multiple generation editing reduces the quality of the image and sound—dramatically in the case of VHS—it is merely an inconvenience in editing, since the purpose of the process is to produce an Edit Decision List (EDL) for use in the on-line edit.

A proprietary computer system, such as Editrack, Cuedos, Shotlister or Turbotrace, will be used which 'remembers' the source material of the original shots when an edit is re-recorded to another generation. Linear off-line editing is rapidly being superseded by non-linear off-line editing.

Editing the Picture: 3

Non-linear off-line editing

In recent years, the post production process has been revolutionised by non-linear systems such as Avid, Lightworks, Montage and D-Vision. Here, the pictures, either from the master videotape or from an off-line copy are stored on hard disks inside an editing computer. Once the material is in the computer it can be edited and subsequently re-arranged almost instantly in the same way as text in a word processor. Because of this flexibility these systems are becoming increasingly popular, but the disadvantage is the limit to the amount of pictures and sound which can be stored in them at any one time. High budget feature films get round this problem by adding large numbers of extra hard disks to the system, but the editor on the average production is frequently faced, after cutting a sequence, with the problem of having to delete material that he might want to refer to in order to free disk space to store fresh material. For this reason an off-line quality VTR is normally included in a non-linear editing suite for retransferring material back into the system.

All the systems allow editors the option of reasonable quality pictures tying up considerable amounts of storage, or poor quality ones using less storage. Unlike a linear off-line editing system, however, the pictures do not degrade through subsequent edits.

At the end of the non-linear off-line, an EDL is produced for use in the on-line.

On-line editing

This is the expensive process of editing the master, at which, if there has been an off-line edit, most of the shots are cut together automatically, using the off-line's EDL. However, complicated digital video effects, captions and graphics, key and matt effects would be rehearsed and inserted as part of a creative process. In the case of productions using a large number of these effects, the off-line edit would have been a very basic affair and may not have taken place at all. Many producers, if they can afford it, particularly in broadcast television magazine programmes, will go directly from shooting to an on-line edit.

Non-conventional film editing

With the advent of timecode systems in film shooting, Aatoncode, Arricode, Keycode, etc., and the increasing use of timecode to synchronise the film camera to its associated sound recorder, quarter-inch tape or DAT, it is now possible to perform what, in effect, is an off-line video edit on a film. By means of a suitable computer system, the negative film rushes are synchronised on a roll of videotape, probably of off-line quality. Technically, rather than creatively, this videotape can be regarded as the first edit of the material and it is therefore possible, after the off-line 'video' edit has taken place to trace the timecodes of the original film which passed through the camera. These 'edit decisions' can be passed to

161

a negative cutter who then cuts the actual film negative, which is therefore available for printing in the conventional way. This method of working is becoming increasingly popular, partly because the flexibility of non-linear editing systems can appeal to film editors and also because, by transferring the film negative directly to videotape, considerable money is saved by not printing any of the film rushes.

A far better way, if a film production is made entirely for television, is to transfer the film negative directly to digital videotape, perform a non-linear off-line edit, and then conform the mastered videotape as if the programme had been originated on tape. The two main advantages to this way of working are:

> considerably better quality by leaving out the film printing process and the precious master material is copied at an early stage in the production—if the master videotape of the rushes is lost or damaged you can always go back to the film.

The disadvantage is that as the negative is not cut, a conventional film print cannot be made for optical projection.

PAs take note!
Because there are no rolls of film or boxes of videotape in non-linear off-line editing, it is absolutely essential on any type of production that the PA's shooting log, i.e. shot lists, continuity sheets etc. is *totally accurate*.

If, at the end of the day, you find you haven't written down the camera roll, put 'don't know' rather than guess. Inaccurate information can, in extreme cases, mean that the editor is unable to access a particular shot from the computer, even though it has been recorded into the machine. Unlike assistant film editors, computers never say 'I can't find exactly what you asked for, try this'.

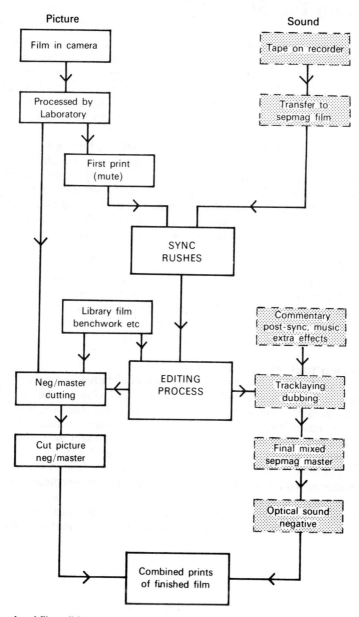

Picture

- Film in camera
- Processed by Laboratory
- First print (mute)

Sound

- Tape on recorder
- Transfer to sepmag film

SYNC RUSHES

Library film benchwork etc

EDITING PROCESS

Commentary post-sync, music extra effects

Neg/master cutting

Tracklaying dubbing

Cut picture neg/master

Final mixed sepmag master

Optical sound negative

Combined prints of finished film

Conventional film editing
This is what happens to the film after a day's shooting.
In television, the show print is usually mute and is shown in conjunction with the final sepmag track on a double-head machine.

Editing the Sound: 1

It is important to remember that you cannot record sound or mix it together in a film cutting-room. All the sound required—actuality sound shot on location, library sound effects, other sound specially recorded for the production such as Foley effects, music, commentary, etc.—have to be transferred to magnetic sound film of the same gauge as the picture and laid on to tracks separated by blank film called 'spacing' in sync with the picture.

A dubbing cue sheet is prepared by the editor and the dubbing mixer works from this in mixing the various tracks together in a dubbing theatre to eventually form one master sound track. In the course of this operation, various pre-mixes may be made—dialogue, music and effects, etc.—either because more tracks have been laid than can be handled by the equipment in the dubbing theatre in one pass, or, particularly in the case of M & E tracks, because they may be required for foreign sales.

Videotape
Very often in videotape editing all the various sounds required in the finished production are mixed together by the editor at the same time as he edits the picture. Alternatively, tracks can be laid as with film, except that they are re-recorded by the editor in sync with the master tape on to video or audio tapes with matching timecodes.

Non-linear track-laying
After a programme has been edited by conventional film or videotape editing, it is possible for any or all or the track-laying to take place in a non-linear fashion on a variety of systems, of which Audiofile is perhaps the best known. These systems manipulate the sound in much the same way as a non-linear editing system handles pictures and sound, except in this case the sound is of on-line, or broadcast quality. Some non-linear editing systems have options, such as Lightworks Pro-sound, which allow the editor to work with off-line picture quality and on-line sound quality. Alternatively, it is possible to perform a complete off-line track-lay within a non-linear editing system and then, using the resultant EDL, conform the track-lay either in an on-line editing suite or in a system such as Audiofile.

DUBBING CUE SHEET

REEL NO. _DR 12345_

PRODUCTION _" LONG WEEKEND "_

PROG NO. _1/ABC F 985J_

PRODUCER/DIRECTOR _JOHN FELTON_

EDITOR _MARK LEWIS_

ACTION	1 A+B/M+S/MONO	2 A+B/M+S/MONO	3 A+B/M+S/MONO	4 A+B/M+S/MONO	MISC
000 TITLE SEQ		0000 OPENING TITLE MUSIC			
0028 GENERAL SHOTS SEAFRONT AND TOWN	0028 DIAL. + FX PREMIX	0028			
			0041 COMM ...JUST POSSIBLE 0058		
			0065 COMM ...AND AGAIN 0103		
0104 INT. DISCO	0104	0104 ROCK MUSIC		0104 ↓ 0138 POSS. DISCO-TYPE CHAT UNDER MUSIC?	
			0110 COMM ...AND DISCOS 0136		
0137 PEACEFUL COUNTRYSIDE	0137 FX PREMIX	(nat.end) 0138			
			0142 COMM		

JBH/DUBCUE #2 21/07/93

Editing the Sound: 2

If you're confused
Let's take the example of a door slam that was not recorded on location. First of all the sound itself has to be acquired, either as a library effect or by the rather obvious process of taking a sound recordist to a door and getting someone to slam it!

In conventional film track-laying, the sound is transferred on to 16mm magnetic film (in the case of a 16mm film production) and placed with other sound effects in a roll of spacing so that the sound of the door slam is exactly the same number of feet and frames away from the start mark at the beginning of the roll as the picture of the door being actually slammed.

With linear videotape editing, let us assume that the actual point of the door slam on the picture is at a timecode of 10.01.02.03. The sound of the slam is then re-recorded on a timecoded tape at 10.01.02.03.

With a non-linear track-laying system, once the door slam has been transferred into the system, the editor simply tells the computer to play it at 10.01.02.03. It doesn't need to be re-recorded on to anything, and one sound copy of one door being slammed can be played out as many times as is required.

What does a PA do?
The PA might be required to find library effects, tapes, discs or CDs for the dubbing session. During the final run-through of the master sound track the PA should time any music for copyright details.

Pre-dubbing script
The script is merely an accurate shot list of the edited production, with the relevant film footage or edited tape timecode given on one side. In the case of timecode, it is often more convenient for the hours and frames to be left off, i.e. 10.11.09.18 is written as 11.10. You should write the footage or timecode from the start of the first shot. This script can be used for writing commentary.

Commentary script
If the PA is required to type a script for the commentary it should be in treble spacing, with the footage or timecode on the left-hand side. A wide margin should be left for notes and alterations. Sentences should never be carried forward from one page to another.

Pre-dubbing post-production script

SHOT NO:	DESCRIPTION	FOOTAGE
1	W/A harbour	000
2	LS boat about to dock	032
3	MS man walking down gangplank	041
	(steps onto land) S/I MAIN TITLE	047
4	Man walking onto railway platform	064
5	LS train coming into station	092
6	Train stops, man gets in	119
7	L/A train departing	143
8	Int. train. SYNC to cam	168
9	B/W Library shot. LMS steam train	182

Commentary script

FOOTAGE   COMMENTARY

119   I'm now going to travel on the wildest railway line in Great Britain, part

of which was originally called the Duke of Sutherland's railway. When

the line was built, it had a private station at Dunrobin to serve the Duke's

castle. That station's still there.

(to clear by 143 for train effects)

167

Copyright: 1

Copyright is the protection the law (in the UK, the Copyright Designs and Patents Act 1988) gives to a person or organisation for something that they have created.

Protection
The protection the law gives ensures that no one other than the copyright owners have the right to use the work unless authorised by the owner, who is thus able to charge a fee. It is particularly important that permission is obtained before any material is used in a production if that material is copyright.

Simple copyright
In its simplest form the material could be a literary work, a painting, a photograph, a musical composition or a recorded musical performance. But such things as records, illustrated books, films and videos involve multiple copyright.

Multiple copyright
Let us take the example of a film. Copyright in a film belongs to the film maker. However, it is not as simple as that, as the film distributor may have bought some or all the rights in that film. For example, the film *Heavens Above* was a Boulting Brothers production, distributed to cinemas by British Lion Films and available on 16 mm for non-theatrical use from Filmbank Distributors Ltd. Before using an excerpt of this film in another production one has to trace which organisation is in a position to give the necessary permission.

Suppose a brasswear manufacturer had bought the rights to include an excerpt of *Heavens Above* in a corporate video called *Brass* because they had made the altar candlesticks seen in the film. No one, including the brasswear manufacturer, has the right to use that sequence in any other video without negotiating fresh rights from British Lion, even though they may have copied the sequence from *Brass*. The brasswear manufacturer does not own the copyright of the sequence, only a licence permitting him to use it in *Brass*.

Copyright extends for 50 years from the film being registered under Part III of the Cinematographic Films Act 1950 (UK), or, if not registered, 50 years from when it was first published, i.e. when copies were sold or hired to the public.

Music copyright
A commercial recording involves multiple copyright: the composers'; the performers';
the recording company's.

Copyright: 2

Photographs
Copyright extends for 50 years from when the photograph was first published.

Artistic works
Copyright extends for 50 years from the year of the author's death.

Music
The composer's copyright extends for 50 years from the death of the composer. But be careful, because if the work has since been re-arranged by someone, the arranger enjoys the same copyright protection as the original composer. With gramophone records and CDs the copyright resides with the manufacturers of the record or CD for 50 years from when the recording was first published.

The following organisations in the UK deal with the copyright use of music in a programme:

The Performing Right Society,
29–33 Berners Street,
London WIP 4AA
Telephone: 071 580 5544

The Mechanical Copyright
Protection Society,
Elgar House,
41 Streatham High Street,
London SW16 1ER
Telephone: 081 769 4400

Phonographic Performance Ltd.,
14–22 Ganton Street,
London W1V 1LB
Telephone: 071 437 0311/6

Literary works
Copyright extends for 50 years from the end of the year in which the author died.

One final word. If, say, an artist is commissioned to do a work, then the copyright of that work may well rest with the person or organisation who commissioned it and not with the artist.

Take care
Suppose you are shooting an item on the availability of English magazines in a Spanish holiday resort. The wide shot of the street contains a magazine-stand, which although including all sorts of copyright material such as advertising hoardings, would not be regarded as a breach of copyright. Similarly, a tighter shot of the stand itself showing a whole number of magazines would not infringe any rights. But a close-up of a

single magazine could infringe copyright. It might be necessary, as a courtesy, to contact the copyright holder and ask permission before using such a shot.

NOTE: All references to copyright made here apply to the law in force in the UK, although in general they apply in the US also.

Literary works
Copyright extends for 50 years from the end of the year in which the author died.

Post Production Script

Throughout this book you have probably been aware that a large part of your job has been to record events. These may be events that have not yet taken place, as with the shooting schedule; those that are in the process of taking place, as with the continuity report sheet or the shot list; or events that have already taken place, as with the post production script.

This script is the written account of the production. It contains *everything* that is in the finished production and, more than that, it comprises the source of everything you have used, whether it be specially shot for the production, film archive material, music or photographs.

Use of the script

The post production script has a twofold use:

1. It is used for costing purposes. As it contains the original source of everything used in the production, including details of the artists, it makes the job of paying the correct people relatively simple.
2. It is used for archive purposes. If anyone in future years wishes to use any part of the material it will be easy for them to find all the requisite details from this script.

What the script contains

The script contains all the action (in the form of a shot list), together with the dialogue, against the film footage or timecode reference.

Library film

If library film is used, this is shown and the copyright source given.

Music

Full details of all music must be given—whether it is specially composed or taken from a commercial recording, in which case details of the maker, the performers, the composer and the record publisher must be given.

Photographs

The original source of all photographs must be given.

POST PRODUCTION SCRIPT EXAMPLE

FOOTAGE	ACTION	SOUND
000	W/A harbour	Natural f/x and music *'The Three Elizabeths Suite'*, comp. & cond. Eric Coates Orchestra: The New Symphony DECCA LK4056
032	LS boat about to dock	
041	MS man walking down gangplank	
047	(He steps on to land) S/I MAIN TITLE	
064	Man walking on to railway platform	Music ends 064 - train f/x only
092	LS train into stn.	
119	Train stops. Man gets in	<u>COMM</u>: I'm now going to travel on the wildest railway line in Great Britain, part of which was originally called the Duke of Sutherland's railway. When the line was built it had a private station at Dunrobin to serve the Duke's castle. That station is still there.
143	L/A train departing	F/X train departing
168	Int. train. Man speaking	<u>SYNC</u>: There's even a buffet car on this amazing service which runs three times a day but never on Sundays. Before the war they served kippers.
182	B/W LIBRARY SHOT. LMS steam train near Altnabreac (private collection: R. McTavish, 4 Shaw St., Lochinver, Sutherland)	Music *'The Four Centuries Suite'*, comp. & cond. Eric Coates Orchestra. The New Symphony DECCA LK 4056

173

Film and Videotape Formats: 1

Film
If the production is being shot on film it will be shot on 16mm, Super 16mm or 35mm film stock. 35mm film stock is the best quality and is used for feature films, TV commercials and pop promotions. Until a year or so ago, 16mm was the normal format for television production, but this is increasingly giving way to Super 16mm. The advantage of Super 16mm is that it is a wide screen image. A production shot on Super 16mm will have a longer 'shelf life' because it is the right shape for widescreen television. However only the centre part of the image is likely to be transmitted on today's television screens.

Videotape
There are in existence quite a bewildering array of videotape formats and sizes. New products are arriving on the market so quickly that the comprehensive list that follows may be incomplete by the time you read it.

Analogue and digital
The simplest way to understand the difference between analogue and digital is to think of a domestic audio cassette tape and a compact disc. The audio cassette is an analogue recording and the compact disc is a digital one. The CD gives spectacularly better quality. Digital video recording gives a similar improvement when compared to analogue video recording.

Most of the new formats developed in the last few years have been digital rather than analogue. Apart from higher quality, the main advantage of digital VTRs is that there is minimal loss of quality when pictures are copied, as in editing, compared to analogue VTRs. In addition, many digital VTRs offer a facility known as read-before-write, which means you can replay an existing picture off a VTR at the same time as replacing it with a fresh one. This is especially useful in editing applications involving complex multi-layering, but is best explained by the example of adding name superimpositions to an already edited programme using ONE machine only. The image is replayed (read) off the machine, combined with a character generator in a vision mixer and re-recorded (written) on the same place on the tape.

Although to get the maximum advantage from a digital VTR it would have to be used in an entirely digital environment, i.e. digital camera, digital vision mixer in the edit suite and so on, there is still a spectacular improvement when digital VTRs are used in an otherwise analogue environment.

A further complication arises as it is possible to work component or composite in either analogue or digital.

Component and composite
A composite signal, such as from a 1" VTR, utilises one wire to carry all the

picture information. A component signal, such as Betacam SP, uses three wires, one for the black and white information and two for the colour information.

While component is a better way to process television signals, giving better quality pictures, it is not a convenient way to distribute those pictures to the viewer. All the national terrestrial transmissions are composite systems—PAL in the UK and most of Europe, SECAM in France and Russia, and NTSC in America and Japan.

Film and Videotape Formats: 2

Videotape formats

All the following recording formats are linear, i.e. they involve magnetic tape moving continuously past a rotating head drum. Non-linear video recordings are an entirely new development and mostly found in off-line editing systems such as AVID and Lightworks. Systems like AVID NEWS-CUTTER and HEAVYWORKS however give broadcast quality pictures from non-linear sources.

Analogue

1. *2″ Quadruplex videotape (composite).* This was invented by Ampex in 1956 and was the standard tape used for broadcasting throughout the world for over twenty years. Nowadays 2″ VT would only be encountered by the PA when working with archive material. Most large broadcasting organisations retain a machine for replaying archive tapes and there are specialist facility houses who undertake transfers of Quad tape to modern formats.

A 2″ machine will not give a recognizable picture until it has run up to speed, which takes ten seconds, and one cannot see any sort of picture when spooling fast. If 2″ material is to be incorporated into an on-line edit, it must first be transferred to a modern format since today's editing systems cannot control 2″ VTRs.

Many programmes recorded on 2″ will not have any timecode, and if the PA is required to log shots from a 2″ programme without timecode, she should ask the tape operator to set the machine's tape counter to zero at an agreed point on the tape and make a note of that point. She can then use the machine's counter to log the material that needs to be transferred. When the material is transferred to a modern format, timecode should be requested for the new recording and this timecode should be noted by the PA in her editing notes.

2. *1″ Videotape (composite).* 1″ videotape superseded 2″ throughout the broadcast industry but is now itself being superseded by digital formats. 1″ videotape allows for still frame and slow speed replay together with a picture of sorts when spooling. Most 1″ machines have timecode.

2″ and 1″ videotapes are wound on open spools but all other videotape is contained in cassettes.

3. *¾″ High Band U-Matic (composite).* High Band U-Matic spearheaded the change from film to tape for news gathering. Marginally regarded as broadcast quality it has now largely been superseded for broadcast purposes. It is, however, much used in lower budget corporate productions and as an off-line editing format. SP U-Matic is an improved version of High Band and, like High Band, the machines may or may not have timecode.

4. 3¼″ *Low Band U-Matic (composite)*. Never regarded as broadcast quality in the UK, the cassettes are identical to High Band cassettes. High Band and Low band formats are not interchangeable although cassettes of either format can be played back in the machines of the other format for viewing purposes only. Either way round, the picture is in black and white. Some U-Matic machines, however, are switchable between High Band and Low Band.

Low Band is likely to be found in off-line editing systems and, being a robust format, in continuously repeating playback systems found in exhibitions, etc. Timecode is not standard on Low Band machines but may have been incorporated in individual systems by, for example, sacrificing an audio track on the tape.

Film and Videotape Formats: 3

5. *½" Betacam and Betacam SP (component)*. Betacam SP is a vastly-improved development of the original Sony Betacam format, and has virtually become the world-wide medium for single-camera acquisition. Depending which system, NTSC, PAL, etc. is in use, cassettes holding up to thirty-six minutes of material can be contained in a recorder mounted directly on the back of the camera itself. While Betacam cassettes can be replayed in Betacam SP machinery, Betacam SP cassettes can only be replayed in the older Betacam VTRs if these have been specially modified.

Betacam SP is a very high quality analogue component format but is not particularly robust and can suffer from drop-out. Timecode is standard on all Betacam and Betacam SP machines.

6. *½" M2 (component)*. Introduced as a rival format to Betacam, it never found quite the same popularity as Betacam SP and although some major television companies in the UK adopted it as a replacement for 1", it is now more likely to be encountered in professional, rather than broadcast, environments. Timecode is standard on M2 machines.

7. *½" Super VHS*. S-VHS achieves a compromise between composite (using one wire) and component (three wires) by using two wires, one to carry the luminance (black and white) part of the picture, and the other to carry the chrominance (colour) part of the picture. It is used by some organisations for news-gathering and will be found in low-budget corporate productions. It is also a high-end domestic format. Some machines have timecode.

8. *HI-Video 8 (8mm wide)*. A technically similar system to S-VHS, using different width tape. Hi−8 would be used in similar situations to S-VHS. Some machines have timecode.

9. *Video-8 (8mm wide) (composite)*. Mainly found as a domestic camcorder format.

10. *½" Betamax (composite)*. Almost obsolete domestic video format.

11. *½" VHS (composite)*. Domestic video format used world-wide. Also used extensively in television for viewing purposes and low quality off-line editing.

Digital

The following digital formats are either available or projected. All digital VTRs have timecode as standard and at least four audio tracks.

1. *¾" D1 (Digital component)*. Developed by Sony, this is the highest quality VTR format in current use. It is used in applications such as title sequences where very many tape generations are required since the high quality makes these generations imperceptible. It is also used for television commercials post produced on videotape, following a high quality digital telecine transfer from 35mm negative film. D1 VTRs are extremely expensive which is why they are not generally used in broadcast work.

2. *¾" DCT (Digital component)*. Standing for Digital Component Technology, DCT is claimed by Ampex to be the industry's first practical digital component production system available from one manufacturer. It is not widely encountered in the UK.

178

3. *D2 (Digital composite)*. Marketed by Sony and Ampex, D2 looked set to be the standard replacement for 1″ tape throughout the industry. The arrival of D3 has, however, tended to divide the market.

4. *½″ D3 (Digital composite)*. Developed by Panasonic, this format offers similar quality and features as D2 but on a narrower width tape. A D3 camcorder is available. This format has been adopted by some companies, including the BBC, as a replacement for 1″.

5. *½″ D5 (Digital component)*. Another Panasonic format, D5 is a component version of D3. D5 machines can handle pictures in the widescreen 16 x 9 ratio and can also replay D3 tapes. In the UK, Channel 4 has adopted D5 as its transmission format.

6. *½″ Digital Betacam (component)*. A digital version of Betacam which can also handle 16 x 9 widescreen pictures. Some machines can replay analogue Betacam SP and Betacam tapes, however no digital Betacam machines can record in analogue Betacam SP.

7. *¼″ DVC (Digital video cassette)*. At the time of writing it is understood that all the major domestic VCR manufacturers are co-operating on a component system using ¼″ tape, which will give four-and-a-half hours of recording time on a tape only a little longer than an audio cassette.

Although this format is primarily intended for consumer use, the provision of two channels of digital audio at full broadcast quality and the likelihood that the picture quality will approach that of analogue Beta SP means that there is no doubt that it will come into professional use.

Film Running Times

FEET	TIME 35mm	TIME 16mm	FEET	TIME 35mm	TIME 16mm	FEET	TIME 35mm	TIME 16mm
1	0.6	1.6	59	37.8	1m 34.4	117	1m 14.9	3m 07.2
2	1.3	3.2	60	38.4	1m 36.0	118	1m 15.5	3m 08.8
3	1.9	4.8	61	39.0	1m 37.6	119	1m 16.2	3m 10.4
4	2.6	6.4	62	39.7	1m 39.2	120	1m 16.8	3m 12.0
5	3.2	8.0	63	40.3	1m 40.8	121	1m 17.4	3m 13.6
6	3.8	9.6	64	41.0	1m 42.4	122	1m 18.1	3m 15.2
7	4.5	11.2	65	41.6	1m 44.0	123	1m 18.7	3m 16.8
8	5.1	12.8	66	42.3	1m 45.6	124	1m 19.4	3m 18.4
9	5.8	14.4	67	42.9	1m 47.2	125	1m 20.0	3m 20.0
10	6.4	16.0	68	43.5	1m 48.8	126	1m 20.7	3m 21.6
11	7.0	17.6	69	44.2	1m 50.4	127	1m 21.3	3m 23.2
12	7.7	19.2	70	44.8	1m 52.0	128	1m 21.9	3m 24.8
13	8.3	20.8	71	45.5	1m 53.6	129	1m 22.6	3m 26.4
14	9.0	22.4	72	46.1	1m 55.2	130	1m 23.2	3m 28.0
15	9.6	24.0	73	46.7	1m 56.8	131	1m 23.8	3m 29.6
16	10.2	25.6	74	47.4	1m 58.4	132	1m 24.5	3m 31.2
17	10.9	27.2	75	48.0	2m 00.0	133	1m 25.1	3m 32.8
18	11.5	28.8	76	48.6	2m 01.0	134	1m 25.8	3m 34.4
19	12.2	30.4	77	49.3	2m 03.2	135	1m 26.4	3m 36.0
20	12.8	32.0	78	49.9	2m 04.8	136	1m 27.0	3m 37.6
21	13.4	33.6	79	50.6	2m 06.4	137	1m 27.7	3m 39.2
22	14.1	35.2	80	51.2	2m 08.0	138	1m 28.3	3m 40.8
23	14.7	36.8	81	51.9	2m 09.6	139	1m 29.0	3m 42.2
24	15.4	38.4	82	52.5	2m 11.2	140	1m 29.6	3m 44.0
25	16.0	40.0	83	53.1	2m 12.8	141	1m 30.2	3m 45.6
26	16.7	41.6	84	53.8	2m 14.4	142	1m 30.9	3m 47.2
27	17.3	43.2	85	54.4	2m 16.0	143	1m 31.5	3m 48.8
28	17.9	44.8	86	55.0	2m 17.6	144	1m 32.2	3m 50.4
29	18.6	46.4	87	55.7	2m 19.2	145	1m 32.8	3m 52.0
30	19.2	48.0	88	56.3	2m 20.8	146	1m 33.4	3m 53.6
31	19.8	49.6	89	57.0	2m 22.4	147	1m 34.1	3m 55.2
32	20.5	51.2	90	57.6	2m 24.0	148	1m 34.7	3m 56.8
33	21.1	52.8	91	58.2	2m 25.6	149	1m 35.4	3m 58.4
34	21.8	54.4	92	58.9	2m 27.2	150	1m 36.0	4m 00.0
35	22.4	56.0	93	59.5	2m 28.8	151	1m 36.6	4m 01.6
36	23.0	57.6	94	1m 00.2	2m 30.4	152	1m 37.3	4m 03.2
37	23.7	59.2	95	1m 00.8	2m 32.0	153	1m 37.9	4m 04.8
38	24.3	1m 00.8	96	1m 01.4	2m 33.6	154	1m 38.6	4m 06.4
39	25.0	1m 02.4	97	1m 02.1	2m 35.2	155	1m 39.2	4m 08.0
40	25.6	1m 04.0	98	1m 02.8	2m 36.8	156	1m 39.8	4m 09.6
41	26.2	1m 05.6	99	1m 03.4	2m 38.4	157	1m 40.5	4m 11.2
42	26.9	1m 07.2	100	1m 04.0	2m 40.0	158	1m 41.1	4m 12.8
43	27.5	1m 08.8	101	1m 04.6	2m 41.6	159	1m 41.8	4m 14.4
44	28.2	1m 10.4	102	1m 05.3	2m 43.2	160	1m 42.4	4m 16.0
45	28.8	1m 12.0	103	1m 05.9	2m 44.8	161	1m 43.0	4m 17.6
46	29.4	1m 13.6	104	1m 06.6	2m 46.4	162	1m 43.7	4m 19.2
47	30.1	1m 15.2	105	1m 07.2	2m 48.0	163	1m 44.3	4m 20.8
48	30.7	1m 16.8	106	1m 07.8	2m 49.6	164	1m 45.0	4m 22.4
49	31.4	1m 18.4	107	1m 08.5	2m 51.2	165	1m 45.6	4m 24.0
50	32.0	1m 20.0	108	1m 09.1	2m 52.8	166	1m 46.3	4m 25.6
51	32.6	1m 21.6	109	1m 09.8	2m 54.4	167	1m 46.9	4m 27.2
52	33.3	1m 23.2	110	1m 10.4	2m 56.0	168	1m 47.5	4m 28.8
53	33.9	1m 24.8	111	1m 11.0	2m 57.6	169	1m 48.2	4m 30.4
54	34.6	1m 26.4	112	1m 11.7	2m 59.2	170	1m 48.8	4m 32.0
55	35.2	1m 28.0	113	1m 12.3	3m 00.8	171	1m 49.5	4m 33.6
56	35.8	1m 29.6	114	1m 13.0	3m 02.4	172	1m 50.1	4m 35.2
57	36.5	1m 32.2	115	1m 13.6	3m 04.0	173	1m 50.7	4m 36.8
58	37.1	1m 32.8	116	1m 14.2	3m 05.6	174	1m 51.4	4m 38.4

FEET	TIME 35mm	16mm	FEET	TIME 35mm	16mm	FEET	TIME 35mm	16mm
175	1m 52.0	4m 40.0	236	2m 31.0	6m 17.6	297	3m 10.1	7m 55.2
176	1m 52.6	4m 41.6	237	2m 31.7	6m 19.2	298	3m 10.8	7m 56.8
177	1m 53.3	4m 43.2	238	2m 32.3	6m 20.8	299	3m 11.4	7m 58.4
178	1m 53.9	4m 44.8	239	2m 33.0	6m 22.4	300	3m 12.0	8m 00.0
179	1m 54.6	4m 46.6	240	2m 33.6	6m 24.0	301	3m 12.6	8m 01.6
180	1m 55.2	4m 48.0	241	2m 34.2	6m 25.6	302	3m 13.3	8m 03.2
181	1m 55.9	4m 49.6	242	2m 34.9	6m 27.2	303	3m 13.9	8m 04.8
182	1m 56.5	4m 51.2	243	2m 35.5	6m 28.8	304	3m 14.6	8m 06.4
183	1m 57.1	4m 52.8	244	2m 36.2	6m 30.4	305	3m 15.2	8m 08.0
184	1m 57.8	4m 54.4	245	2m 36.8	6m 32.0	306	3m 15.8	8m 09.6
185	1m 58.4	4m 56.0	246	2m 37.4	6m 33.6	307	3m 16.5	8m 11.2
186	1m 59.0	4m 57.6	247	2m 38.1	6m 35.2	308	3m 17.1	8m 12.8
187	1m 59.7	4m 59.2	248	2m 38.7	6m 36.8	309	3m 17.8	8m 14.4
188	2m 00.3	5m 00.8	249	2m 39.4	6m 38.4	310	3m 18.4	8m 16.0
189	2m 01.0	5m 02.4	250	2m 40.0	6m 40.0	311	3m 19.0	8m 17.6
190	2m 01.6	5m 04.0	251	2m 40.6	6m 41.6	312	3m 19.7	8m 19.2
191	2m 02.2	5m 05.6	252	2m 41.3	6m 43.2	313	3m 20.3	8m 20.8
192	2m 02.9	5m 07.2	253	2m 41.9	6m 44.8	314	3m 21.0	8m 22.4
193	2m 03.5	5m 08.8	254	2m 42.6	6m 46.4	315	3m 21.6	8m 24.0
194	2m 04.2	5m 10.4	255	2m 43.2	6m 48.0	316	3m 22.2	8m 25.6
195	2m 04.8	5m 12.0	256	2m 43.8	6m 49.6	317	3m 22.9	8m 27.2
196	2m 05.4	5m 13.6	257	2m 44.5	6m 51.2	318	3m 23.6	8m 28.8
197	2m 06.1	5m 15.2	258	2m 45.1	6m 52.8	319	3m 24.2	8m 30.4
198	2m 06.8	5m 16.8	259	2m 45.8	6m 54.4	320	3m 24.8	8m 32.0
199	2m 07.4	5m 18.4	260	2m 46.4	6m 56.0	321	3m 25.4	8m 33.6
200	2m 08.0	5m 20.0	261	2m 47.0	6m 57.6	322	3m 26.1	8m 35.2
201	2m 08.6	5m 21.6	262	2m 47.7	6m 59.2	323	3m 26.7	8m 36.8
202	2m 09.3	5m 23.2	263	2m 48.3	7m 00.8	324	3m 27.4	8m 38.4
203	2m 09.9	5m 24.8	264	2m 49.0	7m 02.4	325	3m 28.0	8m 40.0
204	2m 10.6	5m 26.4	265	2m 49.6	7m 04.0	326	3m 28.7	8m 41.6
205	2m 11.2	5m 28.0	266	2m 50.3	7m 05.6	327	3m 29.3	8m 43.2
206	2m 11.8	5m 29.6	267	2m 50.9	7m 07.2	328	3m 29.9	8m 44.8
207	2m 12.5	5m 31.2	268	2m 51.5	7m 08.8	329	3m 30.6	8m 46.4
208	2m 13.1	5m 32.8	269	2m 52.2	7m 10.4	330	3m 31.2	8m 48.0
209	2m 13.8	5m 34.4	270	2m 52.8	7m 12.0	331	3m 31.8	8m 49.6
210	2m 14.4	5m 36.0	271	2m 53.5	7m 13.6	332	3m 32.5	8m 51.2
211	2m 15.0	5m 37.6	272	2m 54.1	7m 15.2	333	3m 33.1	8m 52.8
212	2m 15.7	5m 39.2	273	2m 54.7	7m 16.8	334	3m 33.8	8m 54.4
213	2m 16.3	5m 40.8	274	2m 55.4	7m 18.4	335	3m 34.4	8m 56.0
214	2m 17.0	5m 42.4	275	2m 56.0	7m 20.0	336	3m 35.0	8m 57.6
215	2m 17.6	5m 44.0	276	2m 56.6	7m 21.6	337	3m 35.7	8m 59.2
216	2m 18.2	5m 45.6	277	2m 57.3	7m 23.2	338	3m 36.3	9m 00.8
217	2m 18.9	5m 47.2	278	2m 57.9	7m 24.8	339	3m 37.0	9m 02.4
218	2m 19.6	5m 48.8	279	2m 58.6	7m 26.4	340	3m 37.6	9m 04.0
219	2m 20.2	5m 50.4	280	2m 59.2	7m 28.0	341	3m 38.2	9m 05.6
220	2m 20.8	5m 52.0	281	2m 59.9	7m 29.6	342	3m 38.9	9m 07.2
221	2m 21.4	5m 53.6	282	3m 00.5	7m 31.2	343	3m 39.5	9m 08.8
222	2m 22.1	5m 55.2	283	3m 01.1	7m 32.8	344	3m 40.2	9m 10.4
223	2m 22.7	5m 56.8	284	3m 01.8	7m 34.4	345	3m 40.8	9m 12.0
224	2m 23.4	5m 58.4	285	3m 02.4	7m 36.0	346	3m 41.4	9m 13.6
225	2m 24.0	6m 00.0	286	3m 03.0	7m 37.6	347	3m 42.1	9m 15.2
226	2m 24.7	6m 01.6	287	3m 03.7	7m 39.2	348	3m 42.7	9m 16.8
227	2m 25.3	6m 03.2	288	3m 04.3	7m 40.8	349	3m 43.4	9m 18.4
228	2m 25.9	6m 04.8	289	3m 05.0	7m 42.4	350	3m 44.0	9m 20.0
229	2m 26.6	6m 06.4	290	3m 05.6	7m 44.0	351	3m 44.6	9m 21.6
230	2m 27.2	6m 08.0	291	3m 06.2	7m 45.6	352	3m 45.3	9m 23.2
231	2m 27.8	6m 09.6	292	3m 06.9	7m 47.2	353	3m 45.9	9m 24.8
232	2m 28.5	6m 11.2	293	3m 07.5	7m 48.8	354	3m 46.6	9m 26.4
233	2m 29.1	6m 12.8	294	3m 08.2	7m 50.4	355	3m 47.2	9m 28.0
234	2m 29.8	6m 14.4	295	3m 08.8	7m 52.0	356	3m 47.8	9m 29.6
235	2m 30.4	6m 16.0	296	3m 09.4	7m 53.6	357	3m 48.5	9m 31.2

FEET	TIME		FEET	TIME		FEET	TIME	
	35mm	16mm		35mm	16mm		35mm	16mm
358	3m 49.1	9m 32.8	419	4m 28.2	11m 10.4	480	5m 07.2	12m 48.0
359	3m 49.8	9m 34.4	420	4m 28.8	11m 12.0	481	5m 07.9	12m 49.6
360	3m 50.4	9m 36.0	421	4m 29.4	11m 13.6	482	5m 08.5	12m 51.2
361	3m 51.0	9m 37.6	422	4m 30.1	11m 15.2	483	5m 09.1	12m 52.8
362	3m 51.7	9m 39.2	423	4m 30.7	11m 16.8	484	5m 09.8	12m 54.4
363	3m 52.3	9m 40.8	424	4m 31.4	11m 18.4	485	5m 10.4	12m 56.0
364	3m 53.0	9m 42.4	425	4m 32.0	11m 20.0	486	5m 11.0	12m 57.6
365	3m 53.6	9m 44.0	426	4m 32.7	11m 21.6	487	5m 11.7	12m 59.2
366	3m 54.3	9m 45.6	427	4m 33.3	11m 23.2	488	5m 12.3	13m 00.8
367	3m 54.9	9m 47.2	428	4m 33.9	11m 24.8	489	5m 13.0	13m 02.4
368	3m 55.5	9m 48.8	429	4m 34.6	11m 26.4	490	5m 13.6	13m 04.0
369	3m 56.2	9m 50.4	430	4m 35.2	11m 28.0	491	5m 14.2	13m 05.6
370	3m 56.8	9m 52.0	431	4m 35.8	11m 29.6	492	5m 14.9	13m 07.2
371	3m 57.5	9m 53.6	432	4m 36.5	11m 31.2	493	5m 15.5	13m 08.8
372	3m 58.1	9m 55.2	433	4m 37.1	11m 32.8	494	5m 16.2	13m 10.4
373	3m 58.7	9m 56.8	434	4m 37.8	11m 34.4	495	5m 16.8	13m 12.0
374	3m 59.4	9m 58.4	435	4m 38.4	11m 36.0	496	5m 17.4	13m 13.6
375	4m 00.0	10m 00.0	436	4m 39.0	11m 37.6	497	5m 18.1	13m 15.2
376	4m 00.6	10m 01.6	437	4m 39.7	11m 39.2	498	5m 18.8	13m 16.8
377	4m 01.3	10m 03.2	438	4m 40.3	11m 40.8	499	5m 19.4	13m 18.4
378	4m 01.9	10m 04.8	439	4m 41.0	11m 42.4	500	5m 20.0	13m 20.0
379	4m 02.6	10m 06.4	440	4m 41.6	11m 44.0	501	5m 20.6	13m 21.6
380	4m 03.2	10m 08.0	441	4m 42.2	11m 45.6	502	5m 21.3	13m 23.2
381	4m 03.9	10m 09.6	442	4m 42.9	11m 47.2	503	5m 21.9	13m 24.8
382	4m 04.5	10m 11.2	443	4m 43.5	11m 48.8	504	5m 22.6	13m 26.4
383	4m 05.1	10m 12.8	444	4m 44.2	11m 50.4	505	5m 23.2	13m 28.0
384	4m 05.8	10m 14.4	445	4m 44.8	11m 52.0	506	5m 23.8	13m 29.6
385	4m 06.4	10m 16.0	446	4m 45.4	11m 53.6	507	5m 24.5	13m 31.2
386	4m 07.0	10m 17.6	447	4m 46.1	11m 52.2	508	5m 25.1	13m 32.8
387	4m 07.7	10m 19.2	448	4m 46.7	11m 56.8	509	5m 25.8	13m 34.4
388	4m 08.3	10m 20.8	449	4m 47.4	11m 58.4	510	5m 26.4	13m 36.0
389	4m 09.0	10m 22.4	450	4m 48.0	12m 00.0	511	5m 27.0	13m 37.6
390	4m 09.6	10m 24.0	451	4m 48.6	12m 01.6	512	5m 27.7	13m 39.2
391	4m 10.2	10m 25.6	452	4m 49.3	12m 03.2	513	5m 28.3	13m 40.8
392	4m 10.9	10m 27.2	453	4m 49.9	12m 04.8	514	5m 29.0	13m 42.4
393	4m 11.5	10m 28.8	454	4m 50.6	12m 06.4	515	5m 29.6	13m 44.0
394	4m 12.2	10m 30.4	455	4m 51.2	12m 08.0	516	5m 30.2	13m 45.6
395	4m 12.8	10m 32.0	456	4m 51.8	12m 09.6	517	5m 30.9	13m 47.2
396	4m 13.4	10m 33.6	457	4m 52.5	12m 11.2	518	5m 31.6	13m 48.8
397	4m 14.1	10m 35.2	458	4m 53.1	12m 12.8	519	5m 32.2	13m 50.4
398	4m 14.8	10m 36.8	459	4m 53.8	12m 14.4	520	5m 32.8	13m 52.0
399	4m 15.4	10m 38.4	460	4m 54.4	12m 16.0	521	5m 33.4	13m 53.6
400	4m 16.0	10m 40.0	461	4m 55.0	12m 17.6	522	5m 34.1	13m 55.2
401	4m 16.6	10m 41.6	462	4m 55.7	12m 19.2	523	5m 34.7	13m 56.8
402	4m 17.3	10m 43.2	463	4m 56.3	12m 20.8	524	5m 35.4	13m 58.4
403	4m 17.9	10m 44.8	464	4m 57.0	12m 22.4	525	5m 36.0	14m 00.0
404	4m 18.6	10m 46.4	465	4m 57.6	12m 24.0	526	5m 36.7	14m 01.6
405	4m 19.2	10m 48.0	466	4m 58.3	12m 25.6	527	5m 37.3	14m 03.2
406	4m 19.8	10m 49.6	467	4m 58.9	12m 27.2	528	5m 37.9	14m 04.8
407	4m 20.5	10m 51.2	468	4m 59.5	12m 28.8	529	5m 38.6	14m 06.4
408	4m 21.1	10m 52.8	469	5m 00.2	12m 30.4	530	5m 39.2	14m 08.0
409	4m 21.8	10m 54.4	470	5m 00.8	12m 32.0	531	5m 39.8	14m 09.6
410	4m 22.4	10m 56.0	471	5m 01.5	12m 33.6	532	5m 40.5	14m 11.2
411	4m 23.0	10m 57.6	472	5m 02.1	12m 35.2	533	5m 41.1	14m 12.8
412	4m 23.7	10m 59.2	473	5m 02.7	12m 36.8	534	5m 41.8	14m 14.4
413	4m 24.3	11m 00.8	474	5m 03.4	12m 38.4	535	5m 42.4	14m 16.0
414	4m 25.0	11m 02.4	475	5m 04.0	12m 40.0	536	5m 43.0	14m 17.6
415	4m 25.6	11m 04.0	476	5m 04.6	12m 41.6	537	5m 43.7	14m 19.2
416	4m 26.2	11m 05.6	477	5m 05.3	12m 43.2	538	5m 44.3	14m 20.8
417	4m 26.9	11m 07.2	478	5m 05.9	12m 44.8	539	5m 45.0	14m 22.4
418	4m 27.6	11m 08.8	479	5m 06.6	12m 46.4	540	5m 45.6	14m 24.0

FEET	TIME 35mm	16mm	FEET	TIME 35mm	16mm	FEET	TIME 35mm	16mm
541	5m 46.2	14m 25.6	602	6m 25.3	16m 03.2	663	7m 04.3	17m 40.8
542	5m 46.9	14m 27.2	603	6m 25.9	16m 04.8	664	7m 05.0	17m 42.4
543	5m 47.5	14m 28.8	604	6m 26.6	16m 06.4	665	7m 05.6	17m 44.0
544	5m 48.2	14m 30.4	605	6m 27.2	16m 08.0	666	7m 06.3	17m 45.6
545	5m 48.8	14m 32.0	606	6m 27.8	16m 09.6	667	7m 06.9	17m 47.2
546	5m 49.4	14m 33.6	607	6m 28.5	16m 11.2	668	7m 07.5	17m 48.8
547	5m 50.1	14m 35.2	608	6m 29.1	16m 12.8	669	7m 08.2	17m 50.4
548	5m 50.7	14m 36.8	609	6m 29.8	16m 14.4	670	7m 08.8	17m 52.0
549	5m 51.4	14m 38.4	610	6m 30.4	16m 16.0	671	7m 09.5	17m 53.6
550	5m 52.0	14m 40.0	611	6m 31.0	16m 17.6	672	7m 10.1	17m 55.2
551	5m 52.6	14m 41.6	612	6m 31.7	16m 19.2	673	7m 10.7	17m 56.8
552	5m 53.3	14m 43.2	613	6m 32.3	16m 20.8	674	7m 11.4	17m 58.4
553	5m 53.9	14m 44.8	614	6m 33.0	16m 22.4	675	7m 12.0	18m 00.0
554	5m 54.6	14m 46.4	615	6m 33.6	16m 24.0	676	7m 12.6	18m 01.6
555	5m 55.2	14m 48.0	616	6m 34.2	16m 25.6	677	7m 13.3	18m 03.2
556	5m 55.8	14m 49.6	617	6m 34.9	16m 27.2	678	7m 13.9	18m 04.8
557	5m 56.5	14m 51.2	618	6m 35.6	16m 28.8	679	7m 14.6	18m 06.4
558	5m 57.1	14m 52.8	619	6m 36.2	16m 30.4	680	7m 15.2	18m 08.0
559	5m 57.8	14m 54.4	620	6m 36.8	16m 32.0	681	7m 15.9	18m 09.6
560	5m 58.4	14m 56.0	621	6m 37.4	16m 33.6	682	7m 16.5	18m 11.2
561	5m 59.0	14m 57.6	622	6m 38.1	16m 35.2	683	7m 17.1	18m 12.8
562	5m 59.7	14m 59.2	623	6m 38.7	16m 36.8	684	7m 17.8	18m 14.4
563	6m 00.3	15m 00.8	624	6m 39.4	16m 38.4	685	7m 18.4	18m 16.0
564	6m 01.0	15m 02.4	625	6m 40.0	16m 40.0	686	7m 19.0	18m 17.6
565	6m 01.6	15m 04.0	626	6m 40.7	16m 41.6	687	7m 19.7	18m 19.2
566	6m 02.3	15m 05.6	627	6m 41.3	16m 43.2	688	7m 20.3	18m 20.8
567	6m 02.9	15m 07.2	628	6m 41.9	16m 44.8	689	7m 21.0	18m 22.4
568	6m 03.5	15m 08.8	629	6m 42.6	16m 46.4	690	7m 21.6	18m 24.0
569	6m 04.2	15m 10.4	630	6m 43.3	16m 48.0	691	7m 22.2	18m 25.6
570	6m 04.8	15m 12.0	631	6m 43.8	16m 49.6	692	7m 22.9	18m 27.2
571	6m 05.5	15m 13.6	632	6m 44.5	16m 52.2	693	7m 23.5	18m 28.8
572	6m 06.1	15m 15.2	633	6m 45.1	16m 53.8	694	7m 24.2	18m 30.4
573	6m 06.7	15m 16.8	634	6m 45.8	16m 55.4	695	7m 24.8	18m 32.0
574	6m 07.4	15m 18.4	635	6m 46.4	16m 56.0	696	7m 25.4	18m 33.6
575	6m 08.0	15m 20.0	636	6m 47.0	16m 57.6	697	7m 26.1	18m 35.2
576	6m 08.6	15m 21.6	637	6m 47.7	16m 59.2	698	7m 26.8	18m 36.8
577	6m 09.3	15m 23.2	638	6m 48.3	17m 00.8	699	7m 27.4	18m 38.4
578	6m 09.9	15m 24.8	639	6m 49.0	17m 02.4	700	7m 28.0	18m 40.0
579	6m 10.6	15m 26.4	640	6m 49.6	17m 04.0	701	7m 28.6	18m 41.6
580	6m 11.2	15m 28.0	641	6m 50.2	17m 05.6	702	7m 29.3	18m 43.2
581	6m 11.9	15m 29.6	642	6m 50.9	17m 07.2	703	7m 29.9	18m 44.8
582	6m 12.5	15m 31.2	643	6m 51.5	17m 08.8	704	7m 30.6	18m 46.4
583	6m 13.1	15m 32.8	644	6m 52.2	17m 10.4	705	7m 31.2	18m 48.0
584	6m 13.8	15m 34.4	645	6m 52.8	17m 12.0	706	7m 31.8	18m 49.6
585	6m 14.4	15m 36.0	646	6m 53.4	17m 13.6	707	7m 32.5	18m 51.2
586	6m 15.0	15m 37.6	647	6m 54.1	17m 15.2	708	7m 33.1	18m 52.8
587	6m 15.7	15m 39.2	648	6m 54.7	17m 16.8	709	7m 33.8	18m 54.4
588	6m 16.3	15m 40.8	649	6m 55.4	17m 18.4	710	7m 34.4	18m 56.0
589	6m 17.0	15m 42.4	650	6m 56.0	17m 20.0	711	7m 35.0	18m 57.6
590	6m 17.6	15m 44.0	651	6m 56.6	17m 21.6	712	7m 35.7	18m 59.2
591	6m 18.2	15m 45.6	652	6m 57.3	17m 23.2	713	7m 36.6	19m 00.8
592	6m 18.9	15m 47.2	653	6m 57.9	17m 24.8	714	7m 37.0	19m 02.4
593	6m 19.5	15m 48.8	654	6m 58.6	17m 26.4	715	7m 37.6	19m 04.0
594	6m 20.2	15m 50.4	655	6m 59.2	17m 28.0	716	7m 38.2	19m 05.6
595	6m 20.8	15m 52.0	656	6m 59.8	17m 29.6	717	7m 38.9	19m 07.2
596	6m 21.4	15m 53.6	657	7m 00.5	17m 31.2	718	7m 39.6	19m 08.8
597	6m 22.1	15m 55.2	658	7m 01.1	17m 32.8	719	7m 40.2	19m 10.4
598	6m 22.8	15m 56.8	659	7m 01.8	17m 34.4	720	7m 40.8	19m 12.0
599	6m 23.4	15m 58.4	660	7m 02.4	17m 36.0	721	7m 41.4	19m 13.6
600	6m 24.0	16m 00.0	661	7m 03.0	17m 37.6	722	7m 42.1	19m 15.2
601	6m 24.6	16m 01.6	662	7m 03.7	17m 39.2	723	7m 42.7	19m 16.8

FEET	TIME 35mm	16mm	FEET	TIME 35mm	16mm	FEET	TIME 35mm	16mm
724	7m 43.4	19m 18.4	785	8m 22.4	20m 56.0	846	9m 01.4	22m 33.6
725	7m 44.0	19m 20.0	786	8m 23.0	20m 57.6	847	9m 02.1	22m 35.2
726	7m 44.7	19m 21.6	787	8m 23.7	20m 59.2	848	9m 02.7	22m 36.8
727	7m 45.3	19m 23.2	788	8m 24.3	21m 00.8	849	9m 03.4	22m 38.4
728	7m 45.9	19m 24.8	789	8m 25.0	21m 02.4	850	9m 04.0	22m 40.0
729	7m 46.6	19m 26.4	790	8m 25.6	21m 04.0	851	9m 04.6	22m 41.6
730	7m 47.2	19m 28.0	791	8m 26.2	21m 05.6	852	9m 05.3	22m 43.2
731	7m 47.8	19m 29.6	792	8m 26.9	21m 07.2	853	9m 05.9	22m 44.8
732	7m 48.5	19m 31.2	793	8m 27.5	21m 08.8	854	9m 06.6	22m 46.4
733	7m 49.1	19m 32.8	794	8m 28.2	21m 10.4	855	9m 07.2	22m 48.0
734	7m 49.8	19m 34.4	795	8m 28.8	21m 12.0	856	9m 07.8	22m 49.6
735	7m 50.4	19m 36.0	796	8m 29.4	21m 13.6	857	9m 08.5	22m 51.2
736	7m 51.0	19m 37.6	797	8m 30.1	21m 15.2	858	9m 09.1	22m 52.8
737	7m 51.7	19m 39.2	798	8m 30.8	21m 16.8	859	9m 09.8	22m 54.4
738	7m 52.3	19m 40.8	799	8m 31.4	21m 18.4	860	9m 10.4	22m 56.0
739	7m 53.0	19m 42.4	800	8m 32.0	21m 20.0	861	9m 11.0	22m 57.6
740	7m 53.6	19m 44.0	801	8m 32.6	21m 21.6	862	9m 11.7	22m 59.2
741	7m 54.2	19m 45.6	802	8m 33.3	21m 23.2	863	9m 12.3	23m 00.8
742	7m 54.9	19m 47.2	803	8m 33.9	21m 24.8	864	9m 13.0	23m 02.4
743	7m 55.5	19m 48.8	804	8m 34.6	21m 26.4	865	9m 13.6	23m 04.0
744	7m 56.2	19m 50.4	805	8m 35.2	21m 28.0	866	9m 14.3	23m 05.6
745	7m 56.8	19m 52.0	806	8m 35.8	21m 29.6	867	9m 14.9	23m 07.2
746	7m 57.4	19m 53.6	807	8m 36.5	21m 31.2	868	9m 15.5	23m 08.8
747	7m 58.1	19m 55.2	808	8m 37.1	21m 32.8	869	9m 16.2	23m 10.4
748	7m 58.7	19m 56.8	809	8m 37.8	21m 34.4	870	9m 16.9	23m 12.0
749	7m 59.4	19m 58.4	810	8m 38.4	21m 36.0	871	9m 17.5	23m 13.6
750	8m 00.0	20m 00.0	811	8m 39.0	21m 37.6	872	9m 18.1	23m 15.2
751	8m 00.6	20m 01.6	812	8m 39.7	21m 39.2	873	9m 18.7	23m 16.8
752	8m 01.3	20m 03.2	813	8m 40.3	21m 40.8	874	9m 19.4	23m 18.4
753	8m 01.9	20m 04.8	814	8m 41.0	21m 42.4	875	9m 20.0	23m 20.0
754	8m 02.6	20m 06.4	815	8m 41.6	21m 44.0	876	9m 20.6	23m 21.6
755	8m 03.2	20m 08.0	816	8m 42.2	21m 45.6	877	9m 21.3	23m 23.2
756	8m 03.8	20m 09.6	817	8m 42.9	21m 47.2	878	9m 21.9	23m 24.8
757	8m 04.5	20m 11.2	818	8m 43.6	21m 48.8	879	9m 22.6	23m 26.4
758	8m 05.1	20m 12.8	819	8m 44.2	21m 50.4	880	9m 23.2	23m 28.0
759	8m 05.8	20m 14.4	820	8m 44.8	21m 52.0	881	9m 23.9	23m 29.6
760	8m 06.4	20m 16.0	821	8m 45.4	21m 53.6	882	9m 24.5	23m 31.2
761	8m 07.0	20m 17.6	822	8m 46.1	21m 55.2	883	9m 25.1	23m 32.8
762	8m 07.7	20m 19.2	823	8m 46.7	21m 56.8	884	9m 25.8	23m 34.4
763	8m 08.3	20m 20.8	824	8m 47.4	21m 58.4	885	9m 26.4	23m 36.0
764	8m 09.0	20m 22.4	825	8m 48.0	22m 00.0	886	9m 27.0	23m 37.6
765	8m 09.6	20m 24.0	826	8m 48.7	22m 01.6	887	9m 27.7	23m 39.2
766	8m 10.3	24m 25.6	827	8m 49.3	22m 03.2	888	9m 28.3	23m 40.8
767	8m 10.9	20m 27.2	828	8m 49.9	22m 04.8	889	9m 29.0	23m 42.4
768	8m 11.5	20m 28.8	829	8m 50.6	22m 06.4	890	9m 29.6	23m 44.0
769	8m 12.2	20m 30.4	830	8m 51.2	22m 08.0	891	9m 30.2	23m 45.6
770	8m 12.8	20m 32.0	831	8m 51.8	22m 09.6	892	9m 30.9	23m 47.2
771	8m 13.5	20m 33.6	832	8m 52.5	22m 11.2	893	9m 31.5	23m 48.8
772	8m 14.1	20m 35.2	833	8m 53.1	22m 12.8	894	9m 32.2	23m 50.4
773	8m 14.7	20m 36.8	834	8m 53.8	22m 14.4	895	9m 32.8	23m 52.0
774	8m 15.4	20m 38.4	835	8m 54.4	22m 16.0	896	9m 33.4	23m 53.6
775	8m 16.0	20m 40.0	836	8m 55.0	22m 17.6	897	9m 34.1	23m 55.2
776	8m 16.7	20m 41.6	837	8m 55.7	22m 19.2	898	9m 34.8	23m 56.8
777	8m 17.3	20m 43.2	838	8m 56.3	22m 20.8	899	9m 35.4	23m 58.4
778	8m 17.9	20m 44.8	839	8m 57.0	22m 22.4	900	9m 36.0	24m 00.0
779	8m 18.6	20m 46.4	840	8m 57.6	22m 24.0	901	9m 36.6	24m 01.6
780	8m 19.2	20m 48.0	841	8m 58.2	22m 25.6	902	9m 37.3	24m 03.2
781	8m 19.9	20m 49.6	842	8m 58.9	22m 27.2	903	9m 37.9	24m 04.8
782	8m 20.5	20m 51.2	843	8m 59.5	22m 28.8	904	9m 38.6	24m 06.4
783	8m 21.1	20m 52.8	844	9m 00.2	22m 30.4	905	9m 39.2	24m 08.0
784	8m 21.8	20m 54.4	845	9m 00.8	22m 32.0	906	9m 39.8	24m 09.6

FEET	TIME 35mm	16mm	FEET	TIME 35mm	16mm	FEET	TIME 35mm	16mm
907	9m 40.5	24m 11.2	939	10m 01.0	25m 02.4	971	10m 21.5	25m 53.6
908	9m 41.1	24m 12.8	940	10m 01.6	25m 04.0	972	10m 22.1	25m 55.2
909	9m 41.8	24m 14.4	941	10m 02.2	25m 05.6	973	10m 22.7	25m 56.8
910	9m 42.4	24m 16.0	942	10m 02.9	25m 07.2	974	10m 23.4	25m 58.4
911	9m 43.0	24m 17.6	943	10m 03.5	25m 08.8	975	10m 24.0	26m 00.0
912	9m 43.7	24m 19.2	944	10m 04.2	25m 10.4	976	10m 24.6	26m 01.6
913	9m 44.3	24m 20.8	945	10m 04.8	25m 12.0	977	10m 25.3	26m 03.2
914	9m 45.0	24m 22.4	946	10m 05.4	25m 13.6	978	10m 25.9	26m 04.8
915	9m 45.6	24m 24.0	947	10m 06.1	25m 15.2	979	10m 26.6	26m 06.4
916	9m 46.2	24m 25.6	948	10m 06.7	25m 16.8	980	10m 27.2	26m 08.0
917	9m 46.9	24m 27.2	949	10m 07.4	25m 18.4	981	10m 27.9	26m 09.6
918	9m 47.6	24m 28.8	950	10m 08.0	25m 20.0	982	10m 28.5	26m 11.2
919	9m 48.2	24m 30.4	951	10m 08.6	25m 21.6	983	10m 29.1	26m 12.8
920	9m 48.8	24m 32.0	952	10m 09.3	25m 23.3	984	10m 29.8	26m 14.4
921	9m 49.4	24m 33.6	953	10m 09.9	25m 24.8	985	10m 30.4	26m 16.0
922	9m 50.1	24m 35.2	954	10m 10.6	25m 26.4	986	10m 31.0	26m 17.6
923	9m 50.7	24m 36.8	955	10m 11.2	25m 28.0	987	10m 31.7	26m 19.2
924	9m 51.4	24m 38.4	956	10m 11.8	25m 29.6	988	10m 32.3	26m 20.8
925	9m 52.0	24m 40.0	957	10m 12.5	25m 31.2	989	10m 33.0	26m 22.4
926	9m 52.7	24m 41.6	958	10m 13.1	25m 32.8	990	10m 33.6	26m 24.0
927	9m 53.3	24m 43.2	959	10m 13.8	25m 34.4	991	10m 34.2	26m 25.6
928	9m 53.9	24m 44.8	960	10m 14.4	25m 36.0	992	10m 34.9	26m 27.2
929	9m 54.6	24m 46.4	961	10m 15.0	25m 37.6	993	10m 35.5	26m 28.8
930	9m 55.2	24m 48.0	962	10m 15.7	25m 39.2	994	10m 36.2	26m 30.4
931	9m 55.8	24m 49.6	963	10m 16.3	25m 40.8	995	10m 36.8	26m 32.0
932	9m 56.5	24m 51.2	964	10m 17.0	25m 42.4	996	10m 37.4	26m 33.6
933	9m 57.1	24m 52.8	965	10m 17.6	25m 44.0	997	10m 38.1	26m 35.2
934	9m 57.8	24m 54.4	966	10m 18.3	25m 45.6	998	10m 38.8	26m 36.8
935	9m 58.4	24m 56.0	967	10m 18.9	25m 47.2	999	10m 39.4	26m 38.4
936	9m 59.0	24m 57.6	968	10m 19.5	25m 48.8	1000	10m 40.0	26m 40.0
937	9m 59.7	24m 59.2	969	10m 20.2	25m 50.4			
938	10m 00.3	25m 00.8	970	10m 20.9	25m 52.0			

185